Guide to the
RIBA Domestic and Concise Building Contracts 2014

RIBA ⁂ Publishing

Sarah Lupton

Guide to the RIBA Domestic and Concise Building Contracts 2014

© Sarah Lupton, 2015

Published by RIBA Publishing, part of RIBA Enterprises Ltd,
The Old Post Office, St Nicholas Street, Newcastle upon Tyne, NE1 1RH

ISBN 978 1 85946 545 5

Stock code 82645

British Library Cataloguing-in-Publication Data
A catalogue record for this book is available from the British Library.

Publisher: Steven Cross
Commissioning Editor: Fay Gibbons
Project Editor: Richard Blackburn
Designed and Typeset by Academic + Technical, Bristol, UK
Printed and bound by Page Bros, Norwich, UK
Cover design: Kneath Associates
Cover image: Shutterstock: www.shutterstock.com

RIBA Publishing is part of RIBA Enterprises Ltd.
www.ribaenterprises.com

Foreword

In November 2014, the RIBA published two new forms of building contract, responding directly to members' needs for simple, easy-to-read and flexible forms for use on projects that are, so often, less straightforward than those for which the shorter JCT contracts are intended. The new forms have been well received and, by promoting collaborative team working, providing for flexible payments, staged completions and contractors' programmes, and including a plethora of much needed clarifications, they are a very useful addition to the current forms of contract.

Nonetheless it can be stressful trying to understand the implications of the terms of a contract under which you will be acting as the contract administrator, particularly for a busy architect in practice, and even more so when working with a form of contract with which you have little experience. Help is now, thank goodness, at hand and from the pre-eminent contract guide author of her time.

Sarah Lupton's guide responds to the new contracts perfectly: it explains them in simple, easy-to-read language, with hints, tips and pertinent legal back-up in the form of legislation and case reports. It will provide the reassurance to understand the implications – for all members of the team.

The Guide runs in a logical order, covering when to start thinking about which contract to use, how and when to set out and select the contract terms, what to do to ensure continuity of good relationships, what paperwork is needed, and how best to use the clauses to the advantage of both parties. The tricky topics of change management, programme adherence, and even termination are clearly structured and explained – all with a good dollop of advice from an author who is both an active professional and a calm, knowledgeable, experienced and astute adviser.

This is the book that all members of the team should be issued with, from the moment that one of these forms of Contract has been selected, if not before. Clearly explaining what the new contracts involve, this Guide will be equally invaluable for the student, for the professional considering contractual options and for all parties involved in an active contract. Expect it to become well thumbed!

Jane Duncan
RIBA President Elect

About the Author

Professor Sarah Lupton MA, DipArch, LLM, RIBA, CArb is dual qualified in both architecture and law. She is a partner in Lupton Stellakis, architects, and holds a personal chair at Cardiff University. As a specialist in design liability and construction law she also practises as a Chartered Arbitrator, adjudicator and expert witness, and is the author of a number of books, including *Which Contract?* (RIBA Publishing), the highly popular guides to the JCT contracts (RIBA Publishing) and a new edition of *Cornes and Lupton's Design Liability in the Construction Industry* (Wiley-Blackwell).

Contents

Foreword		iii
About the Author		iv
Contents		v
About this Guide		ix
1	**Introduction**	**1**
	Features of the RIBA Building Contracts	1
	Suitability for different procurement routes	2
	Differences between the concise and domestic contracts	3
	Compliance with the Housing Grants, Construction and Regeneration Act 1996 (as amended)	5
	Other differences between the contracts	6
	Use of CBC by a 'consumer'	7
	Comparison with other contracts	9
2	**Forming the contract**	**11**
	Tendering	11
	Tendering procedures	11
	Information that must be included in tender documents	12
	Pre-contract negotiations	19
	Executing the contract	20
	Interpreting the contract and resolving inconsistencies	21
	Definitions	21
	Priority of contract documents	21
	Inconsistencies, errors or omissions	22
3	**Roles and management systems**	**23**
	Role of the contract administrator	23
	Role of the contractor	25
	Completing the works	26
	Contractor's design obligation	26
	Contractor's obligations in respect of subcontracted work	28
	Compliance with statute/health and safety legislation	30
	Role of the client	34
	Customer acting as contract administrator	36
	Management systems	37
	Pre-start meeting	37
	Progress meetings (CBC)	38
	Risks register	38

4	**Project progress**	**39**
	The site: possession	39
	Starting the work	40
	Completion in sections	40
	The contractor's programme	40
	Content of programme	41
	Progress	42
	Updated programmes	42
	Finishing the work	43
	Delay	43
	Revisions to the contract completion date	44
	Applying for a revision of time	46
	Assessment of an application	47
	Final assessment of revisions of time	49
5	**Control of the works**	**51**
	Control of day-to-day activities	51
	'Person-in-charge'	51
	Responsibility for subcontractors	51
	Principal designer	52
	Flow of information	52
	Information to be provided by the contract administrator	52
	Information to be provided by the contractor	53
	Inspection and tests	54
	Inspection	54
	Testing and defective work	55
	Contractor administrator's instructions	55
	Delivery of instructions	56
	Procedure following an instruction	57
	Practical completion	58
	Consequences of practical completion	59
	Use/occupation before practical completion	60
	Non-completion	62
	The defects fixing period	62
6	**Payment and certification**	**65**
	The contract price	66
	Adjustments to the contract price	67
	Certification and payment – concise contract	70
	Advanced payment	70
	Interim payments – monthly certification	70
	Interim payments – payment on practical completion	75
	Interim payments – milestone payments	75
	Payment of interim payments	76
	Final contract price and payment	77
	Certification and payment – DBC	79
	Interim payments – monthly certification	79
	Interim payments – milestone payments	80
	Payment of interim payments	81
	Payments following practical completion	82
	Final contract price and payment	83

Conclusiveness   84
Non-payment and non-certification   84
    Contractor's remedy if no certificate issued   84
    Contractor's remedy if payment not made   85

7   **Insurance**   **87**
Liability   87
Indemnity   89
Insurance   89
    Professional indemnity insurance   91

8   **Termination**   **93**
Termination by the client   93
    Abandoning the works   94
    Failing to proceed regularly and diligently   94
    Consistently failing to comply with instructions 95
    Material breach of contract   95
Termination by the contractor   96
    Failure to pay the contractor when payment is due 96
    Material breach of contract   96
Termination by either party   96
    Insolvency or bankruptcy   97
    Frustration   97
Procedure for terminating   97
Consequences of termination   98
    Payment   98
    Access to the site and security   100

9   **Dispute handling and resolution**   **101**
Mediation   101
Adjudication   102
    The adjudication process   103
    Challenging an adjudicator's decision   104
Arbitration   104

References   **107**

Further Reading   **109**

Clause index   **111**

Subject index   **115**

About this Guide

The new RIBA Concise Building Contract 2015 and RIBA Domestic Building Contract 2014 are welcome additions to the range of standard contracts available for use on small and medium-scale projects that follow a traditional procurement route. They have been developed in response to feedback from RIBA members, and have been designed to be clear, flexible and suited to the needs of consultants and their clients.

This Guide is intended to assist users of the new contracts. It assumes no prior knowledge of the contracts or of construction contract law. As well as examining the provisions of the contracts, it also covers related legal topics, such as relevant legislation, contract formation and dispute resolution.

The first chapter of the Guide discusses the situations where it may be appropriate to use these contracts, particularly in the context of the Housing Grants, Construction and Regeneration Act 1996 and consumer protection legislation, and outlines the contracts' key provisions. Detailed tables compare the contracts with each other and with other standard building contracts. The second chapter examines information that should be included in the tender package, and the formation and execution of the contract. The Guide then introduces the roles of the customer and employer (referred to as the 'client'), the contractor and the contract administrator, presenting tables of their duties and powers. The remaining chapters look in detail at programming and revisions of time, control of the works, certification and payment, insurance, termination and dispute resolution.

The contracts contain a wide variety of useful provisions, some of which may be new to users, such as the collaborative working (advance warnings, joint resolution of delay, proposals for improvements and cost savings) and project management provisions (pre-start meeting and progress meetings), and the time-bar clauses in relation to contractor claims. They also include some useful optional provisions, for example for milestone payments, for a contractor programme (with penalties for non-provision), for required specialists (for whom the contractor takes entire responsibly), for contractor design, with a 'fit for purpose' liability option, for new building warranties and for a risk register. All of these are discussed in the Guide.

As with any new contract, the RIBA Building Contracts do not yet have a proven track record. However, assurances can be given as to their legal robustness and usability in practice; a number of potential users, experts and lawyers were asked to review drafts of the contracts and their comments acted upon, and the contracts have been written in plain English to ensure that their provisions will be easily understood by any user. Although there is no case law directly related to the RIBA Building Contracts, it is still possible to analyse their provisions by analogy with existing case law on similar contracts. A comparative method has therefore been used throughout this Guide, both as a means of explaining key provisions and to justify any comments and interpretation of various clauses.

When first using one of the new RIBA Building Contracts, the user may need to take some expert advice, particularly in complex areas such as insurance.

In summary, the new RIBA Building Contracts offer attractive alternatives to existing contracts in that they are relatively short and easy to read, and yet contain a range of innovative features not found in other standard contracts. As such they are likely to prove a popular choice with their intended users.

1 Introduction

1.1   In November 2014 the Royal Institute of British Architects (RIBA) launched two new building contracts: The RIBA Concise Building Contract 2014 and the RIBA Domestic Building Contract 2014. These are not the first contracts to be published by the RIBA. In fact, the very first UK standard form of building contract was published by the RIBA jointly with the Institute of Builders and the National Federation of Building Trades Employers of Great Britain and Northern Ireland in 1902 (and cost one shilling!). In 1931 the Joint Contracts Tribunal (JCT) was formed, and the drafting and publishing of the contract became a pan-industry project. It continued to be called the RIBA Standard Conditions of Contract until 1963, when it was republished as the JCT Standard Form of Building Contract.

1.2   The JCT now publishes an extensive suite of contracts. However, the RIBA identified a need, through feedback from its members, for short, easy-to-read and flexible contracts that would be suitable for less straightforward projects than those catered for by the JCT Home Owner contracts and that could be used on domestic and commercial works. It has therefore published two new forms, designed for use in conjunction with RIBA's architect/consultant appointment agreements, the RIBA Domestic Project Agreement 2010 (2012 revision) and the RIBA Concise Agreement 2010 (2012 revision).[1]

Features of the RIBA Building Contracts

1.3   According to its guidance notes, the RIBA Concise Building Contract 2014 (CBC) is intended for use on 'all types of simple commercial building work'. It can be used in both the private and public sectors, as it incudes optional provisions dealing with official secrets, transparency, discrimination and bribery as normally required by public sector clients (item BB and clause A9). The RIBA Domestic Building Contract 2014 (DBC), as its name suggests, is intended for domestic work, including renovations, extensions, maintenance and new buildings. Its guidance notes state that this is limited to 'work done to the Customer's home', i.e. to contracts that fall under the 'residential occupier exception' in the Housing Grants, Construction and Regeneration Act 1996, as amended by Part 8 of the Local Democracy, Economic Development and Construction Act 2009 (hereafter referred to as the Housing Grants Act) (see para. 1.14). DBC is endorsed by the HomeOwners Alliance.

1.4   Both contracts are available in hard copy and electronic formats, and can be purchased from: www.ribacontracts.com.

1.5   Key features of the RIBA Building Contracts are:

- collaboration provisions: advance warnings, joint resolution of delay, proposals for improvements and costs savings;
- management provisions: pre-start meeting (both contracts), progress meetings (CBC only);

---

[1]   The RIBA intends to update its suite of architect/consultant agreements in the near future; the new agreements will also align with the new RIBA Building Contracts.

- flexible payment options;

- provision for contractor design, with a 'fit for purpose' liability option;

- optional provisions for a contractor programme, with penalties for non-provision;

- optional provisions for client-selected suppliers and subcontractors;

- mechanisms for dealing with changes to the project which allow for agreement and include specified timescales;

- option for commencement and completion in stages;

- terms compliant with the Unfair Terms in Consumer Contracts Regulations 1999 for consumer clients;

- guidance notes on use and completion are included.

1.6   All of these features are discussed at various points throughout this Guide.

Suitability for different procurement routes

1.7   Both of the RIBA Building Contracts are suitable for projects that are procured on what is normally referred to as a 'traditional' procurement route, i.e. one where the client engages at least one, and possibly several, firms of consultants to prepare a design and complete full technical documentation before the project is tendered to contractors. With traditional procurement, it is common that some parts of the design are completed by the contractor or by specialist subcontractors (this version is sometimes referred to as 'traditional plus contractor design'). The RIBA Building Contracts would be suitable in this situation, with either the main contractor or a subcontractor (client selected, if required) undertaking the design. Traditional procurement is widely used – it was applied in around 76 per cent of projects undertaken in the UK in 2010, which accounted for 41 per cent of the total value (RICS and Davis Langdon, 2012).

1.8   The RIBA Building Contracts are not intended for use in design and build procurement, i.e. where the contractor is responsible for both the design and the construction of the project. It would be possible to adapt them for this use, but there would still be a need for a contract document giving full details of the design requirements, and for an appointed contract administrator (not usual in design and build contracts). In addition, all the provisions relating to design would require amendment, and more detailed provisions regarding submission and approval of the developing design would have to be added.

1.9   The contracts are also not intended to be used with management contracting or construction management arrangements (where the project is tendered as a series of packages to separate firms, with work progressing on a rolling programme basis after the first package is let). Management procurement arrangements are normally used on very large projects; however, on a smaller scale, clients who wish to manage projects themselves sometimes adopt a similar system and engage a number of separate companies (often referred to as 'separate trades'). The RIBA Building Contracts might be suitable for some of the larger work packages, but thought should be given as to how all the separate contracts are to be co-ordinated. This is not an easy task, and the apparent savings achieved by cutting out the main contractor's markup may be more than offset by the amount of time the client has to spend managing the process, or by the additional fees charged by consultants if they undertake this role.

1.10   However, within the traditional plus contractor design route, the RIBA Building Contracts would be suitable for a wide range of projects, from very small-scale alterations and refurbishment works, to moderately sized projects relating to existing or new buildings. As mentioned above, CBC's guidance notes refer to 'simple' work, and generally it is the level of complexity, rather than the value, that should be the key determinant in the choice of contract.

1.11   Although the RIBA Building Contracts have many useful features (see para. 1.5), which mean they are flexible, some of the provisions lack the detail to be found in larger contracts. Examples are those relating to design submission and approval procedures and insurance clauses. Other provisions commonly found in larger contracts are not included, such as fluctuations clauses and contractor bonds – if these are required then other standard contracts should be considered.

Differences between the concise and domestic contracts

1.12   The key difference between the two contracts is that, as noted above, CBC complies with the Housing Grants Act, whereas DBC does not. This results in significant differences in the payment and dispute resolution terms (see paras. 1.14–1.19). In addition, each version contains some features that are not included in the other.

1.13   A full comparison of the differences between the two versions is set out in Table 1.1. The key differences in this list are set in bold, with those relating to the Housing Grants Act underlined.

Table 1.1 Key differences between CBC and DBC

Title/Subject	CBC clauses	DBC clauses	Notes
Main clauses			
Obligation	1.1	1.1	same
Carrying out the works	1.2–1.5	1.2–1.5	same
Subcontracting	1.6–1.7	1.6–1.7	same
Employer/customer	2.1–2.4	2.1–2.4	same
Pre-start meeting	3.1	3.1	same
Advance warning and joint resolution of delay	3.2–3.3	3.2–3.3	same
Improvements and cost savings	3.4–3.5	3.4–3.5	same
Progress meetings	3.6	–	**not included**
Assignment and health and safety	4.1–4.3	4.1–4.3	same
Architect/contract administrator	5.1–5.3	5.1–5.3	same
Instructions	5.4–5.8	5.4	same
Contractor notification regarding instructions	5.9–5.10	–	not included
Contractor failure to comply with an instruction	5.11	5.8	same
Change to works instructions	5.12–5.13	5.9–5.10	same
Inconsistency in the contract documents	5.14–5.15	5.11–5.12	same

Table 1.1 Key differences between CBC and DBC – Continued

Title/Subject	CBC clauses	DBC clauses	Notes
Contractor's liability	6.1	6.1	same
Employer/customer's liability	6.2	6.2	same
Insurance (heading missing in DBC)	6.3	6.3	same
Evidence	6.4	6.4	same
Insurance backed guarantee		6.5	DBC only
Payment to employer/customer	6.5	6.6	same
Interim payments	7.1–7.10	7.1–7.7	different text
Final contract price	7.11–7.12	7.8	different text
Final payment (heading missing in DBC)	7.13–7.15	7.9	different text
Suspension	8.1	8.1	same
Right to interest	8.2	–	not included
Entitlement to costs	8.3	8.2	same
Employer's/customer's risks	9.1–9.2	9.1–9.2	same
Contractor's risks	9.3–9.4	9.3–9.4	same
Force majeure	9.5–9.8	9.5–9.8	same
Revision of time	9.9–9.12	9.9–9.12	same
Additional payment	9.13–9.14	9.13–9.14	same
Practical completion	9.15–9.16	9.15–9.16	same
Early use and partial possession	9.17–9.18	9.17–9.18	same
Liquidated damages	10.1	10.1	same
Defects fixing period	10.2–10.3	10.2–10.3	same
Priority of contract documents	11.1–11.2	11.1–11.2	same
Governing law, etc.	11.3–11.9	11.3–11.9	same
Communication and notices	11.10–11.12	11.10–11.12	same
Calculation of periods	11.13	11.13	same
Cancellation by customer		12.1–12.2	DBC only
Termination by the employer/customer	12.1–12.2	12.3–12.4	same
Termination by the contractor	12.3–12.4	12.5–12.6	same
Termination due to insolvency	12.5	12.7	same
Termination by either party	12.6	12.8	same
Payment on termination	12.7–12.8	12.9–12.10	same
Payment on termination	12.9	12.11	different text
Access to site and security on termination	12.10	12.12	same

Table 1.1 Key differences between CBC and DBC – Continued

Dispute resolution	13	13	different text: adjudication is a right in CBC, optional in DBC
Optional clauses			
Programme	A1	A1	**CBC additionally requires updates before progress meetings**
Programme and progress meetings	A1.2–A1.3	–	**not included**
Contractor design	A2	A2	same
Completion in sections	A3	A3	same
Payment on completion of the works and milestone payments	A4	A4	different text: DBC allows for milestone payment only
Advanced payment	A5	–	not included
Evidence of ability to pay the contract price	A6	–	not included
Required specialists	A7	A5	same
Customer acting as contract administrator		A6	**DBC only**
New building warranty		A7	DBC only
Collateral warranty/third party rights agreement	A8	–	not included
Public sector clauses	A9	–	not included
Risks register	A10	A8	same
Rules for valuation of revision of time and additional payment	A11	A9	same

Compliance with the Housing Grants, Construction and Regeneration Act 1996 (as amended)

1.14   The Housing Grants Act applies to most construction contracts.[2] It requires that construction contracts include specific terms relating to adjudication and payment:

- the right to stage payments;
- the right to notice of the amount to be paid;
- the right to suspend work for non-payment;
- the right to take any dispute arising out of the contract to adjudication.

1.15   If the parties fail to include these provisions in their contract, the Act will imply terms to provide these rights (section 114) by means of the Scheme for Construction Contracts (England and Wales) Regulations 1998.

---

[2]   The term 'construction contract' is given a wide definition, and includes contracts for services only or for building work, and applies to a large range of types of work, including demolition works, services installations and repair work, as well as new buildings, extensions and alterations.

1.16   However, there is an important exception; it does not apply to projects where one of the parties is a 'residential occupier'. The residential occupier exception applies to projects for which the primary purpose is beneficial use by the client as a residence (section 106). This would include buildings that the client is occupying or intending to occupy as its main residence, and might also include a second home if the client is the main user and there is no intention to use it as a holiday let (*Westfields* v *Lewis*). However, work to buildings in the grounds of a residence that will not be lived in by the customer, or work to divide a property into flats where only one flat will be retained by the customer, will not fall under the exception. Similarly, work on other residential properties, for example for landlords, local authorities or housing associations, will usually be covered by the Act.

1.17   CBC 2014 contains provisions that comply with the Housing Grants Act and can therefore be used on any project, including those to which the Act applies, whereas DBC 2014 can only be used for 'residential occupier' projects.

1.18   The effect of this is that the payment terms differ considerably between the concise and the domestic versions. In brief, CBC sets up 'Due Dates', allows the contractor to apply for payment not later than ten days before each due date, requires the contract administrator to certify payment within five days of the due date, and requires the employer to pay within 14 days after the due date. If the employer wishes to pay less than the amount certified, it must issue a 'Pay Less Notice' not later than five days before the final date for payment. In DBC, however, the procedure is much simpler; there are no due dates, the contract administrator issues certificates at monthly intervals, the contractor then issues an invoice to the customer, and the customer pays with 14 days of receipt of the invoice. If the customer decides to make a deduction, it explains the reason for the deduction at the time of making the payment (these provisions are explained in full in Chapter 6).

1.19   As CBC's additional terms introduce considerable complications, it would be unwise to use CBC on projects with a residential occupier, who may find the provisions onerous; the guidance notes emphasise that it 'is **not** suitable for non-commercial work'. As well as the practicality of operating the provisions, there is also a risk that some of the terms may be caught by the Unfair Terms in Consumer Contracts Regulations 1999 if not individually negotiated with the client prior to entering into the contract (see para. 1.25).

Other differences between the contracts

1.20   Compliance with the Housing Grants Act is obviously a key difference between the contracts, however there are other important differences. For example, DBC has an option whereby the customer may elect to act as contract administrator (cl A6). It also allows for the customer to require an insurance backed guarantee covering the contract price (cl 6.5), or a new building warranty (cl A7, see para. 2.21). Although insurance backed guarantees are normally used in domestic projects, they can be provided for commercial work. If an employer using the CBC requires one, a new clause will need to be introduced to the contract.

1.21   There are also several provisions in CBC that are not included in DBC, such as the requirements to hold progress meetings and for the contractor to submit an updated programme in advance of the meeting. This omission seems a great pity, as both features, particularly the regular programme updates, are useful on any type of project. It may be that the drafters felt this would overcomplicate a domestic project, but as such projects can be significant in size and complexity, and as these features are not to be found in JCT contracts, it may be an opportunity lost.

1.22   CBC includes optional clauses that provide for advance payment, for the employer to provide evidence of ability to pay the contract price, and for the right to interest on unpaid amounts. These are also omitted from DBC, which seems logical in a domestic context (especially the latter, as the Late Payment of Commercial Debts (Interest) Act 1998 would not apply to domestic contracts). CBC also includes an optional clause dealing with collateral warranties and third party rights (see para. 2.28). These are usually provided to future purchasers or tenants, or to funders. DBC does not provide for collateral warranties as such warranties do not normally arise in a domestic context. In the unlikely event that the customer's bank requires a warranty, then an additional provision would need to be introduced.

Use of CBC by a 'consumer'

1.23   The law affords special protection to persons who enter into a contract as a 'consumer', through what is referred to collectively as consumer protection legislation.

1.24   It is important to note that the definitions of 'residential occupier' and 'consumer' are not the same. The usual definition of consumer in relevant legislation is a person who is acting 'for purposes which are outside his trade, business or profession'. This could include, for example, a wealthy business tycoon who is developing a country estate with several large residences for members of his family; in other words, it is not confined to small projects, or to those undertaking work on their own home, or to residential occupiers. Therefore, although a residential occupier will always be a consumer (i.e. the customer in DBC will be a consumer), there may also be situations where a consumer is not a residential occupier, in which case CBC would be the appropriate choice.

1.25   A key piece of legislation affecting consumers is the Unfair Terms in Consumer Contracts Regulations 1999. Essentially, these provide that terms that significantly affect the balance of rights between the parties, and which are not individually negotiated, may be deemed unfair and therefore void.

Unfair Terms in Consumer Contracts Regulations 1999

These Regulations only apply to terms in contracts between a seller of goods or supplier of goods and services and a consumer, and only where the terms have not been individually negotiated (this would generally include all standard contracts). A consumer is defined as a person who, in making a contract, is acting 'for purposes which are outside his trade, business or profession' (section 3(1)). An 'unfair term' is any term that causes a significant imbalance in the parties' rights to the detriment of the consumer, and the regulations state that any such term will not be binding on the consumer. An indicative list of terms is given in Schedule 2 to the regulations and includes, for example, any term 'excluding or hindering the consumer's right to take legal action or exercise any other legal remedy, particularly by requiring the consumer to take disputes exclusively to arbitration'.

1.26   Both of the RIBA Building Contracts have been drafted with the intention that they should be fair. The language is certainly clear and should be relatively easy for a lay client to understand. Nevertheless, it is always wise for an architect advising a customer on procurement matters to go through the terms with any client, ideally before the project is sent out to tender, and certainly before the tender is accepted and the contract formed. They should also ensure that all terms are explained carefully to 'consumer' clients, in order that they can be considered to have been individually negotiated. It is important, for example, that if the arbitration option is selected, or if any other amendments are made which could be seen as limiting the client's rights, these have been explained and discussed. It would also be important that, where CBC is being used with a consumer, the payment provisions are explained, as some of these provisions may be considered onerous.

1.27   Similarly, DBC includes the right of the customer to cancel the contract within 14 days of signing it (cl 12.1), something required by the Consumer Contracts (Information, Cancellation and Additional Charges) Regulations 2013. This right is not included in CBC. Where CBC is being used with a consumer, it might be wise to add this provision, or to make it clear that the employer would have this right in any event under the Regulations.

1.28   Finally, those dealing with consumers should look out for an important change in consumer protection legislation: the Consumer Rights Act 2015 has recently been passed by Parliament, although at the time of writing, the date of coming into force is not yet known. It will replace much of the existing law on consumer protection.

Consumer Rights Act 2015

The Act aims to summarise in one place the obligations of traders in relation to consumers for most types of contract, together with consumers' rights against traders. It includes provisions covering unfair terms that are similar to those discussed above. It also includes additional protection for the consumer. A key clause in the Act is as follows:

50 Information about the trader or service to be binding

(1) Every contract to supply a service is to be treated as including as a term of the contract anything that is said or written to the consumer, by or on behalf of the trader, about the trader or the service, if—

(a) it is taken into account by the consumer when deciding to enter into the contract, or
(b) it is taken into account by the consumer when making any decision about the service after entering into the contract.

The key point to note is that the consumer may rely on anything spoken or written to it as if it were a term of the contract, which would include any statements made about the cost, timescale, accommodation, specification and performance of a finished building or the services to be provided. This would mean that any additional undertakings given by the contractor, for example during the tender period, would effectively become part of the contract between the parties. Other provisions in the Act that are of relevance are the right to a price reduction (section 56) or to require repeat performance (section 55) if the consumer is unhappy with the work that has been carried out.

Comparison with other contracts

1.29   The RIBA Building Contracts are significantly different to others that are currently available. One obvious feature is their length; they are shorter than most other contracts available for the intended scale of work, with the exception of the JCT Home Owner contracts (HO/O), and possibly the NEC3 Engineering and Construction Short Contract (NEC3) (with which they are comparable in length). However, they offer considerably more features, including a large number of optional clauses, than are available in competing contracts (the NEC3 Short Contract does not have the optional clauses that are contained in the full version). In fact, the RIBA contracts even offer some features that are not available in much larger contracts, as can be seen in Table 1.2.

Table 1.2 Comparison with other building contracts

	JCT Standard Building Contract 2011 (SBC11)	JCT Intermediate Building Contract with contractor's design (ICD)	JCT Minor Works Contract With Contractor's Design 2011 (MWD11)	JCT Building Contract and Consultancy Agreement for a Home Owner/Occupier (HO/O)	NEC3 Engineering and Construction Short Contract	RIBA Concise Building Contract 2014	RIBA Domestic Building Contract 2014
Progress meetings						✓	
Contractor design	✓	✓	✓		✓	option	option
Employer/customer selected subcontractors	not for design	✓				option	option
Contractor's design submission	✓						
Professional indemnity insurance	✓	✓				✓	✓
Programme	✓				option	option	option
Sectional completion	✓	✓				✓	✓
Partial possession	✓	✓				✓	✓
Listed items	✓	✓					
Variation quotation	✓			✓		✓	✓
Loss/expense	✓	✓	limited	limited	✓	✓	✓
Variation and revision of time rules	variation	variation				option	option
Customer cancellation							✓
Insurance backed guarantee							option
Risks register						option	option

1.30   In summary, the new RIBA Building Contracts offer attractive alternatives to existing contracts in that they are relatively short and easy to read, and yet contain a range of innovative features not found in other contracts. This will make them appealing both to architects and their clients and to contractors.

1.31   One of the drawbacks of new contracts is that, unlike longstanding ones, they are not tried and tested. Their practical application has not been put to the test, and the exact meanings of clauses have not been established in the courts. Some assurances can, however, be given regarding the RIBA Building Contracts. Drafts of the contracts were subject to extensive comments by potential users, by experts and by lawyers, and the contracts are written in plain English so that the intended meaning is generally clear. It is possible in some instances to use analogy with equivalent clauses in other contracts, where the meaning and effect have been discussed in court cases, as a means of interpreting the provisions of the new contracts. A comparative method has therefore been used throughout this Guide, both as a means of explaining key provisions to readers who might be more familiar with other contracts, and to justify comments and interpretation of various provisions.

2 Forming the contract

2.1   As noted in the previous chapter, the RIBA Building Contracts are most likely to be used on projects that follow a traditional procurement route, with possibly a degree of contractor and/or subcontractor design input. The decision to use a RIBA Building Contract will normally be made at an early stage in the project, and certainly by the end of RIBA Plan of Work Stage 3,[1] as it affects issues such as the distribution of design responsibility between consultants and the contractor and the degree to which the client can control which subcontractors may be used.

2.2   Between the selection of the contract and the start of work on site, various processes will need to be undertaken: preparing the tender package and tendering; post-tender negotiations and appointment of the contractor; and formal execution of the contract. There will also be a pre-contract meeting, at which various matters are agreed. This chapter outlines these processes and considers particular matters that arise in relation to the RIBA Building Contracts.

Tendering

2.3   It can't be overemphasised that, when a project is sent out to tender, the contractor should be given full and detailed information regarding the project requirements. This is the principal means by which a designer can ensure that the quality of detailing, finish and workmanship will reach the required standards. If the project is not to be fully designed by the client's consultants, the contractor will require full information about the design that it is to provide, including any performance specifications.

2.4   Most importantly, full details should be given about which standard building contract is to be used, the particular conditions to be applied and any special terms. Consultants tend to focus on the design, technical and budget aspects, but clarity on the terms of the contract is just as important to the tenderer as these will affect the tender price. It is not advisable to introduce matters such as special contract provisions after a long tendering period, or after negotiations on price have been concluded: if they are not acceptable to the contractor, it will be in a very strong bargaining position. Furthermore, if contract documents are never formally executed, the details included in the tender package may subsequently form the basis of the contract between the parties (see para. 2.33).

Tendering procedures

2.5   Normally, for projects of the scale for which the RIBA Building Contracts are intended, one of two methods will be used: competitive tendering with a small number of contractors,

---

[1]   The RIBA Plan of Work 2013 sets out key stages in a building project; an overview of the Plan of Work can be downloaded at: www.ribaplanofwork.com.

or negotiation with a single contractor. The client's consultants will normally suggest which contractors should be considered, but the client may have worked successfully with firms before, or may have been recommended a firm by others, in which case those firms can also be included.

2.6   With competitive tendering, all tenderers should, of course, be sent identical information, so that they are competing on an equal basis. The tenders are then examined by the client's consultants, and normally the lowest priced one is accepted. With negotiated tendering, only the identified contractor will submit a tender, which is usually subject to discussion before acceptance.

2.7   An alternative method is a two-stage process: initially, a small number of contractors are asked to tender on the basis of less than complete information. Negotiations will then take place with the one that makes the most attractive submission. This method is frequently used where the contractor is expected to have a significant design input but the client wishes to approve this design before entering into the main contract; the selected contractor can then work with the client's consultants in finalising the design. It is unlikely that a two-stage tender would be used for a project based on one of the RIBA Building Contracts, but, if it is, an agreement will be needed to cover the contractor's liability to the client for its contribution to the design process.

2.8   Full guidance on tendering procedures can be found in the *NBS Guide to Tendering: For Construction Projects* (Finch, 2011), JCT's *Tendering Practice Note 2012* (JCT, 2012) and the *RIBA Job Book* (Ostime, 2013), but the most important aspect for the purposes of this chapter is *what to include in the tender documents*.

Information that must be included in tender documents

2.9   The information included in the tender documents will eventually form the basis of the contract itself. Therefore, clearly, the tenderers must be told which version of the contract will be used, i.e. CBC or DBC, and whether it will be entered into as a deed or as a simple contract. It must also be clear which of the tender documents are to be priced by the contractor and, if the documents are to have a priority, what that priority is (see para. 2.16 below).

2.10   The identity of the client (including whether it is acting as an individual or a company) must be crystal clear, as should the location of the site and any restrictions on its use and access. In fact, all matters that will later be entered in the 'Contract Details' (located at the front of the two contracts) should be set out in the tender documents. A full list of the items included in the Contract Details is given in Table 2.1.

2.11   Whoever is responsible for arranging the contract (usually the contract administrator) should discuss and agree all of this information with the client prior to tender. Some items are particularly important to raise at an early stage, so that the client has time to consider the matter and take advice, for example insurance and dispute resolution. Most are discussed in detail in other sections of this Guide (see the references in Table 2.1), but a brief introduction to key items in CBC and DBC is given here. There is also helpful information in the guidance notes to the contracts.

Table 2.1 Contract details

CBC	DBC	Subject	Chapter reference
Main items			
A	A	Contractor: name and details	–
B	B	Employer/customer: name and details	–
C	C	Site: address, and whether it will be occupied (may need to refer to a fuller description in another document)	2.13, 4.2
D	D	Architect/contract administrator	2.14
E	E	Other appointments	2.14
F	F	Contract documents, and any priority between them	2.15–2.16
G	not included	How often progress meetings will be held (weekly or monthly)	3.43
H	G	Start and finish dates, working hour restrictions	2.13
I	H	Permission for the contractor to use facilities	2.13
J	I	Liquidated damages (amount per day)	2.17–2.19
K	J	Defects fixing period (default 12 months)	5.46
L	K	Regulatory consents, fees and charges; is the employer/customer or the contractor responsible for these?	2.20, 3.34
M	L	Insurance: contractor	7.15
N	M	Insurance: employer/customer	7.15–7.16
not included	N	Insurance backed guarantee	2.21
O	O	Contract price	6.5-6.8, 6.53
P	not included	First due date (default 30 days after start date) and final date for payment of interim payments (default 14 days after due date)	6.21, 6.48
Q	not included	Final date for payment of the final payment (default 30 days after due date)	6.56
R	not included	Rate of interest (if none given, will be statutory rate)	6.84
S	P	Methods of dispute resolution	2.24–2.25, 9.19
Optional items			
T	Q	Programme: if one is required, and the details	4.8, 4.10, 4.11
U	R	Contractor design and level of liability	2.26–2.27, 3.19, 5.12
V	S	Completion in sections	2.19, 4.7
W1	not included	Payment on completion of the works	6.39
W2	T	Milestone payments	6.41
X	not included	Advanced payment	6.20, 6.33
Y	not included	Evidence of ability to pay the contract price	1.22, 6.86

CBC	DBC	Subject	Chapter reference
Z	U	Required specialists	3.26
not included	V	Whether the customer will be acting as contract administrator	3.37
not included	W	Whether a new building warranty is required	2.21
AA	*not included*	Collateral warranties and third party rights	2.28–2.30
BB	*not included*	Public sector clauses	–
CC	X	Risks register	3.44
DD	Y	Rules for valuation of revisions of time and additional payment	2.31

Table 2.1 title: **Table 2.1 Contract details – Continued**

Whether the contract will be executed as a deed (Agreement: both contracts)

2.12   The decision as to whether the contract will be executed as a deed or as a simple contract (sometimes referred to as 'under hand') will affect the length of time the contractor will be held liable for a breach of the contract. Under the Limitation Act 1980, in the case of a simple contract the time limit for bringing an action for any breach is six years from the date of the breach, and in the case of a contract entered into as a deed it is 12 years. In the latter, the contractor's risk is obviously greater, which may result in higher tender prices.

The site, and whether the building will be occupied (item C: both contracts)

2.13   If the building is to be occupied during construction, this will have significant implications for the contractor's programme and the contract price. It is almost always slower, and more expensive, to work around partial occupation than it is if given a free run of the site. For example, added health, safety and security measures will be needed, such as the provision of protected access routes to the occupied parts. It will not be sufficient to simply indicate 'yes' as required: full details of the planned occupation will be needed (see para. 4.2). As well as indicating occupation, the client is also required to indicate any working hour restrictions (item H/G), and which facilities (electricity, parking, etc.) the contractor may use free of charge (item I/H, cl 1.5).

The contract administrator and other appointments (items D and E, both contracts)

2.14   The names of the firms engaged by the client should be set out, rather than the individuals, although the main point of contact within each firm could be noted. The contract administrator has a key role in the contract (see Chapter 3) and, although identified, is not a party to the contract (cl 11.5). The client may replace the contract administrator with another firm at any time, provided it notifies the contractor (cl 2.2). The contract administrator may not be the only appointment made by the client as, on all but very small projects, the client may also have engaged engineers, quantity surveyors and, possibly, a project manager (the client's right to make such appointments is set out in cl 2.3). The other consultants have no defined role under the building contract, and all communication should be with the contract administrator, but it is sometimes useful for the contractor to be aware of who else is advising the client.

Contract documents (item F, both contracts)

2.15 All the documents that will form the contract bundle should be clearly identified in the tender package. Many of these will have been prepared by the contract administrator and other consultants identified above, however there are some exceptions. If a contractor's design proposal is required, this will also be a contract document, and details of what is required from the contractor at tender stage, and possibly post tender but before the contract is executed, should be set out (this could be referenced here and/or under item U (CBC) or item R (DBC), see para. 2.27 below). The tender documents may include documents prepared by a required specialist, or by other consultants who are not listed.

2.16 Under item F the parties are invited to rank the documents in order of priority. Although this is optional, it would be very sensible to set out a priority in order to avoid arguments later when, as is almost inevitable on most projects, it is discovered that there are inconsistencies between them (note that it would still be possible for a lower priority document to supplement the material in a higher priority document, however it would not be able to contradict it: see *RWE Npower Renewables* v *JN Bentley Ltd*). The two matters to consider are, of course, quality and quantity, and it may be that the ranking is different for both these, e.g. the specification being the key document for defining quality, and the bill of quantities for amount. It will be particularly important, where the contractor is undertaking design, to state whether the employer's requirements for the design (probably included in the specification) or the contractor's design proposals prevail. Effectively this will determine whether, in cases of conflict, the contractor's primary obligation is to meet the specification requirements, or merely to provide what was set out in the contractor's proposals, even if it is later discovered that those proposals do not meet the client's design requirements. Normally a client would prefer the former approach.

Liquidated damages (item J, CBC; item I, DBC)

2.17 The Contract Details require the parties to insert an agreed amount of liquidated damages per day, and the contractor should be informed of the rate at the tender stage. This is an amount to be allowed by the contractor in the event of failure to complete by the date for completion (cl 10.1, note that if the works are divided into sections under optional clause A3, several different rates may be applied).

2.18 The amount must be calculated on the basis of a genuine pre-estimate of the loss likely to be suffered (see *Alfred McAlpine Capital Projects* v *Tilebox*). Provided that the sum is calculated on this basis, it will be recoverable without the need to prove the actual loss suffered, and irrespective of whether the actual loss is significantly less or more than the recoverable sum. In other words, once the rate has been agreed, both parties are bound by it. If 'nil' were to be inserted into the contract particulars then this would preclude the client from claiming any damages at all (see *Temloc* v *Errill*). If no sum were entered, the client may be able to claim general damages, but if this were the intention it would be better to set this out clearly rather than simply to leave the section blank.

2.19   If completion in sections is selected in item V (CBC) or item S (DBC), rates of liquidated damages are required for each section. Note that in this case, the rate should reflect the realistic losses to the client of not getting access to that section, which (unlike partial possession) may not necessarily relate to the floor area of that part. In addition, the professional fees element may not be proportionally reduced.

Alfred McAlpine Capital Projects Ltd v *Tilebox Ltd* [2005] BLR 271

This case contains a useful summary of the law relating to the distinction between liquidated damages and penalties. A JCT Standard Form of Building Contract With Contractor's Design 1998 contained a liquidated damages provision in the sum of £45,000 per week. On the facts, this was a genuine pre-estimate of loss but the actual loss suffered by the developer, Tilebox, was higher. The contractor therefore failed to obtain a declaration that the provision was a penalty. The judge also considered a different (hypothetical) interpretation of the facts whereby it was most unlikely, although just conceivable, that the total weekly loss would be as high as £45,000. In this situation also the judge considered that the provision would not constitute a penalty. In reaching this decision he took into account the fact that the amount of loss was difficult to predict, that the figure was a genuine attempt to estimate losses, that the figure was discussed at the time that the contract was formed and that the parties were, at that time, represented by lawyers.

Temloc Ltd v *Errill Properties* Ltd (1987) 39 BLR 30 (CA)

Temloc entered into a contract with Errill Properties to construct a development near Plymouth. The contract was on a JCT Standard Form of Building Contract 1980 and was in the value of £840,000. '£ nil' was entered in the contract particulars against clause 24·2, liquidated and ascertained damages. Practical completion was certified around six weeks later than the revised date for completion. Temloc brought a claim against Errill Properties for non-payment of some certified amounts, and Errill counterclaimed for damages for late completion. It was held by the court that the effect of '£ nil' was not that the clause should be disregarded (because, for example, it indicated that it had not been possible to assess a rate in advance), but that it had been agreed that no damages would be payable in the event of late completion. Clause 24 is an exhaustive remedy and covers all losses normally attributable to a failure to complete on time. The defendant could not, therefore, fall back on the common law remedy of general damages for breach of contract.

Regulatory consents, fees and charges (item L, CBC; item K, DBC)

2.20   This item asks the parties to indicate who will be responsible for obtaining and paying for 'all planning permissions, Building Regulations approvals and party wall consents'. It will almost always be the case that the client takes responsibility for planning permission, as, unless the project is to be entirely designed by the contractor, which is unlikely, the contractor will not be engaged at the time planning permission is usually sought. The application and any permission issues are normally resolved before going out to tender (the RIBA Plan of Work 2013 notes that 'Planning applications are typically made using the Stage 3 output'). It is possible, however, that the contractor may be required to deal with Building Regulations approval on small projects (by means of a building notice), but for more complex schemes the client will take responsibility. In the case of party wall consents, under the Party Wall etc. Act 1996 the building owner is required to serve notices on all adjoining owners of intended work on or near the site boundaries, and in many cases to appoint surveyors to act on its behalf. It is not possible to simply require the contractor to take on these roles under the building contract; a separate appointment would be needed (with either the contractor or another consultant).

Insurance backed guarantee and new building warranty (items N and W, DBC)

2.21　　These items require the client to set out details of any insurance backed guarantee and/ or new building warranty that will be required. These are similar products; the term 'new building warranty' is used mainly in the context of new homes, whereas 'insurance backed guarantee' can be applied to any type of building work. The National House Building Council's (NHBC) Buildmark scheme is a well-known example of a new building warranty, which covers homes built by NHBC-registered builders. The scheme protects the homeowner against defects for ten years; for the first two years the builder is responsible for putting right any defects, and during the third to tenth years any damage to a home resulting from building defects is covered by an insurance policy. It will also protect the homeowner if the builder goes bankrupt during the course of the work but before completion. Various trade associations offer a range of insurance backed guarantees for domestic and commercial work, for example the Federation of Master Builders' Build Assure scheme, which offers a variety of packages, from simple liquidation cover to ten years' protection against structural damage. If selected, the insurance backed guarantee is to be provided before the start date (cl 6.5).

Contract price (item O, both contracts)

2.22　　Both contracts give two options for setting the contract price: 'Fixed amount' or 'Amount calculated in accordance with the attached schedule of rates'. Item F makes reference to a 'pricing document', so if the fixed price option is selected, the pricing document could be, for example, a fully priced bill of quantities, a schedule of works or a contractor-prepared price breakdown.

2.23　　If 'Amount calculated in accordance with the attached schedule of rates' is selected then a schedule of rates will be included. This will give rates for all types of work envisaged, but not quantities, and will have been prepared by one of the client's consultants. Alternatively, if preferred, the contractor can be asked to prepare the document. The schedule of rates option is useful for projects where it is difficult to assess the likely quantity of work in advance, such as in refurbishment work, where the extent of repair work cannot be ascertained until demolition work is complete. However, there will be no overall contract sum, and it will be difficult to control the overall budget.

Dispute resolution (item S, CBC; item P, DBC)

2.24　　Three forms of dispute resolution are listed in the RIBA Building Contracts: adjudication, mediation and arbitration. In DBC, all three can be selected, or none, or any combination. In CBC, the adjudication option is preselected to ensure that the contract complies with the Housing Grants, Construction and Regeneration Act 1996 (as amended), which requires that all contracts covered by the Act include the right to adjudication. Otherwise, the same applies as for DBC, i.e. the parties may choose any combination.

2.25   It is important to give careful consideration to the choice of dispute resolution options. The differences between them are significant, and it might not be possible for one party to persuade the other to use a different system after a dispute has arisen; this therefore is the only opportunity to determine how matters will be resolved. The different options are discussed in Chapter 9.

Contractor design (item U, CBC; item R, DBC)

2.26   If this item is selected it requires the client to indicate which aspects of the works are to be designed by the contractor. It is important that any aspect, whether a component, element or system (such as services), is delineated with care, as otherwise there may later be disagreements as to who is responsible for interfaces between these and any consultant-designed areas. In addition, the client must indicate whether the contractor's liability is to use reasonable skill and care or that the design is to be 'fit for purpose' (see para. 3.16). It is very important that this entry is not overlooked as the consultant and client must give careful consideration to the issue of liability before sending out the tender documents.

2.27   As well as indicating which parts are to be designed by the contractor, the client is likely to want to stipulate requirements that the designed parts should meet. It would be sensible for the contract administrator to indicate in item U (CBC) or item R (DBC) where these requirements are set out (quite often this will be in the form of a performance specification). In addition, if the contractor is to be required to submit design proposals at tender stage, then this should be made clear in the tender documents, including the nature and extent of the submission required. Similarly, it may be sensible to set out details of any information to be submitted during the works, and the dates by which it will be required (cl A2.2).

Collateral warranty/third party rights (item AA and clause A8, CBC)

2.28   A collateral warranty is a separate contract entered into by the contractor and a third party, typically a purchaser or tenant of the completed building or a project funder. They are particularly important if the contractor is undertaking design, as without one the purchaser, tenant or funder will not be able to recover its losses should the building later develop defects.

2.29   If the employer wishes the contractor to provide such a warranty, this should be made clear in the tender documents (if not the contractor may later agree to provide one, but the employer would not be able to insist that it does). The details of to whom the warranty is to be provided and the form of warranty to be used should also be given. Standard forms of warranty are published by the JCT, although these may require some modification for use with the RIBA Building Contracts. The contractor must execute the warranties as set out within 14 days of being requested to do so by the employer (cl A8.1).

2.30   As an alternative to the use of a collateral warranty, the contract refers to the use of a third party rights agreement. This facility was introduced by the Contracts (Rights of Third Parties) Act 1999. Until this Act came into force, it was a rule of English law that only the two parties to a contract had the right to bring an action to enforce its terms (termed 'privity of contract'). Now, the Act entitles third parties to enforce a right under a contract, where the term in question was intended to provide a benefit to that third party. The third party could be specifically named or it could be an identified class of people. The Act, however, allows for parties to agree that their contract will not be subject to its provisions, and many standard forms adopt this course in order to limit the parties' liabilities. Both RIBA Building Contracts state 'Third parties have no rights under the Contract unless specifically stated in the Contract' (cl 11.6). The concise contract, however, provides for the employer to require the contractor to enter into a 'Third Party Rights Agreement' (cl A8), details of which must be set out in the Contract Details. It should be noted that there are no standard forms of agreement available for use with the RIBA Building Contracts, but it may be possible to adapt the Draft Third Party Rights Schedule published for use with the RIBA Standard Agreement 2010 (2012 revision).

Rules for valuation of revision of time and additional payment (item DD, CBC; item Y, DBC)

2.31   The parties are given the option of agreeing rules to be used for assessing revisions of time, and additional payment. Item DD (CBC) and item Y (DBC) require an indication that rules will apply, but not what these rules will be. Clause A11/A9 then requires the parties to set the rules that will be used, but the contract does not give any details as to when and how this might be done. It would be sensible to indicate what the rules are in the tender documents, or that they will be discussed and agreed at the pre-start meeting. A set of rules commonly adopted is the Society of Construction Law Delay and Disruption Protocol.

Pre-contract negotiations

2.32   Once the tenders have been returned, the next stage is to finalise the agreement with the selected contractor. Ensuring that a contract is in place before construction work begins is a key task of the contract administrator, and to neglect to do so may constitute negligence.

2.33   A contract is formed when an unconditional offer is unconditionally accepted. In the context of a building project, where contractors have been invited to submit competitive tenders, the tenders constitute 'offers' to carry out the work shown in the tender documents for the price tendered. If a tender is accepted unconditionally then a contract will have been formed, and the terms of the contract will be those set out or referred to in the tender documents.

2.34   'Letters of intent' can cloud the picture and should be avoided. If it is possible to accept the tender without qualification then it is better simply to write a letter to that effect, and the contract comes into existence from the moment the letter has been received by the contractor.[2] The effect of a letter expressing an intention to enter into a contract at some point in the future will depend on the wording and circumstances in each case, but it is likely to be of no legal effect. Starting work on such a basis could have disastrous consequences for both parties.

2.35   If there is a period of negotiation before the formal contract is drawn up, careful records should be kept of all matters agreed in order that they can be accurately incorporated into the formal contract documents. These documents should always be prepared as soon as

---

[2]   For a more detailed analysis of the formation of contracts the reader could refer to Furst, Stephen and Vivien Ramsey (eds.), *Keating on Construction Contracts*, 9th edn, (London: Sweet & Maxwell, 2012).

an agreement is reached, and before work commences on site. Failure to execute the documents does not necessarily mean that no contract is in existence, but it can often lead to avoidable arguments about what was agreed.

2.36   The formal contract, once executed, will supersede any conflicting provisions in the accepted tender and will apply retrospectively (*Tameside Metropolitan BC* v *Barlow Securities*).

> *Tameside Metropolitan Borough Council* v *Barlow Securities Group Services Limited* [2001] BLR 113
>
> Under JCT63 Local Authorities, Barlow Securities was contracted to build 106 houses for Tameside. A revised tender was submitted in September 1982 and work started in October 1982. By the time the contract was executed, 80 per cent of building work had been completed, and two certificates of practical completion were issued relating to seven of the houses in December 1983 and January 1994. Practical completion of the last houses was certified in October 1984. The retention was released under an interim certificate in October 1987. Barlow Securities did not submit any final account, although at a meeting in 1988 the final account was discussed. Defects appeared in 1995, and Tameside issued a writ on 9 February 1996. It was agreed between the parties that a binding agreement had been reached before work had started, and the only difference between the agreement and the executed contract was that the contract was under seal (thereby extending the time for bringing a claim from six years to 12 years from any breach of contract). It was found that there was no clear and unequivocal representation by Tameside that it would not rely on its rights in respect of defects. Time began to run in respect of the defects from the dates of practical completion; the first seven houses were therefore time barred. Tameside was not prevented from bringing the claim by failure to issue a final certificate.

Executing the contract

2.37   The RIBA Building Contracts have an 'Agreement' section at the start of the document that is to be signed by both parties. The contracts can be executed as a simple contract (sometimes termed 'under hand' in other contracts) or as a deed; if neither option is selected, the default is a simple contract. It is good practice to sign not only the form itself, but also a copy of all other contract documents, to avoid any arguments about exactly which contract or revisions are included in the contract. In the concise version the employer is given the option of signing as an individual or as a registered company; obviously, if it signs as an individual, the employer will be personally liable to the contractor, whereas as a company its liability will be limited (see *Hamid* v *Francis Bradshaw Partnership*). The option of signing as a company is not offered in DBC, because a company would not fall under the residential occupier exception discussed at paragraph 1.16. Therefore, DBC cannot be used by, for example, a homeowner who wishes to engage a contractor through a company they own.

2.38   Once the tender process is complete and the contract has been signed, the contract administrator is required to give the contractor 'two free copies of the Contract Documents' (cl 5.2.1). This could be in any format the parties agree; there is no need to sign multiple sets, one set should be signed and then copies made. It may be that the contractor would prefer to have one of these as a secure PDF. The original is usually retained by the client, with the contract administrator holding a copy (or the other way round). Any updates are to be issued to the parties promptly (cl 5.2.2); this might happen, for example, if the parties agree an amendment. Any amendment should be in writing and signed by both parties (cl 11.9).

Interpreting the contract and resolving inconsistencies

2.39   The RIBA Building Contracts are clearly laid out and generally easy to understand. However, it should be remembered that the contract between the parties comprises not only the signed contract form, but all the contract documents as well. Even if great care has been taken in preparing the contract documents, there can be unintended conflicts or inconsistencies between them. As a result, it may be difficult to anticipate their combined effect. This section looks at how the contract itself and the package of contract documents are to be interpreted, and how any errors or problems are corrected.

Definitions

2.40   An 'Explanation of Terms' is set out on page 2 of both versions of the contract. Although not described as 'definitions', this is effectively what the explanations are, and in any event they would be used to interpret the meanings of clauses. For example, 'Change to Works Instruction' is described in the concise contract as 'an instruction from the Architect/ Contract Administrator, as described in clauses 5.12 and 5.13, that alters or modifies the design, quality and/or quantity of the Works'. The contract also includes 'defined terms' that are not included in the Explanation of Terms but are defined within the Contract Details. These terms are all capitalised throughout the contract and include, for example, 'Contractor', 'Customer' and 'Start Date'.

2.41   There is also an additional definition to be noted, which occurs within the Contract Conditions themselves: 'a day' is defined in clause 11.13, which explains that in calculating periods of days, all days including weekends should be included, except for public holidays.

Priority of contract documents

2.42   Clause 11.1 states that:

> All parts of the Contract, including the Contract Documents, shall be read together as a whole; however in all circumstances the Agreement, Contract Details and these Contract Conditions shall take precedence over all other Contract Documents.

This means that conflicting provisions in other documents, for example in the preamble to a bill of quantities, will not override the printed conditions. This gives clarity and avoids unintended clashes occurring. However, if the parties wish to agree special terms that differ in any way from the printed conditions, then it will not be sufficient to simply append them in a separate document: amendments must be made to the actual printed contract (note: this is easier to do in the on-line version). This could be done by amending the individual clauses in the printed contract; alternatively, clause 11.1 could be amended by adding a qualification that the revisions set out in another document will take precedence over the printed contract. However, amending standard contracts is unwise without expert advice, as the consequential effects are difficult to predict. Deleting clause 11.1 would be particularly unwise as it may have unintended effects on the contract as a whole.

2.43   Subject to clause 11.1 (i.e. that the printed conditions take precedence), the parties are invited to set out a priority between the other contract documents (cl 11.2). If no priority is set out and a conflict exists that gives rise to a dispute, the adjudicator or court would need to decide, on a balance of probabilities, what the parties' intentions were when they made the contract. This is done objectively, i.e. the adjudicator or court is not interested in the subjective intention, but in what a disinterested bystander would conclude the parties had meant. A contract administrator trying to resolve conflicts in this situation would need to be equally objective; i.e. when looking at the documents as a whole, what, on balance, do they appear to mean?

Inconsistencies, errors or omissions

2.44   The contractor is required to notify the contract administrator if it finds any inconsistencies within or between the contract documents, and/or any instruction, or between the documents and the law (cl 5.14 CBC, cl 5.11 DBC). The contract administrator is required to issue instructions to deal with the issue, and the contractor must take any minimal action needed to comply with the law while awaiting the instruction (cl 5.15 CBC, cl 5.12 DBC). The contract administrator is also given the power under clause 5.4.6 to issue instructions to correct any inconsistency in the contract documents, which would cover issues that have not been notfied. Although expressed as a power, and not a duty (as in JCT forms), it is suggested that the contract administrator should normally take steps to resolve any problems that emerge.

2.45   An instruction to resolve an inconsistency in the contract documents is stated to be a ground for a revision of time, unless the inconsistency is due to a document prepared by the contractor (cl 9.9.6; note: this would cover both inconsistencies within contract documents and inconsistencies between such documents and any others). Other than this, the contract gives no details as to the consequences of such an instruction. Change to works instructions are dealt with separately (cl 5.4.1, see para. 5.23), and the consequences covered in clause 5.12 in CBC and clause 5.9 in DBC, but there is no equivalent clause to deal with any other type of instruction. However, it is quite possible that a clause 5.14 or 5.4.6 instruction will effectively result in a change to the works. For example, if two different documents showed different quantities of brickwork, and the one with the smaller quantity has been ranked higher in priority, then the contract price will be deemed to only allow for the smaller quantity; if more is needed, this is a change to the works. It is suggested that the contract administrator should take a practical approach and treat the correction of an inconsistency as a change unless it falls within the exceptions set out in clause 9.9.6.

3 Roles and management systems

3.1   Contracts perform many functions: they assign risk, they set out obligations and rights, and they can also act as a management tool.

3.2   Generally speaking, where a contract states that an action 'shall' be performed, this indicates an obligation, also referred to as a 'duty'; where it says it 'may' be performed, this is an optional action, referred to as a 'right' or a 'power'. Some of the obligations and rights could be described as core; for example, the contractor's duty to complete the works, the client's duty to pay the contractor, and the contract administrator's right to instruct a change to the works. Some are procedural, for example the obligation to submit a document within a particular time frame. The RIBA Building Contracts additionally have a group of clauses, under the heading 'Collaborative Working', which are aimed at assisting the smooth management of the contract.

3.3   This chapter describes the roles of the contract administrator, the contractor and the client, outlining their duties and rights. It then looks at some of the provisions in the RIBA Building Contracts that aim to ensure best practice management procedures are applied to projects.

Role of the contract administrator

3.4   The contract administrator is appointed by the client, and its duties and liabilities are owed to the client as set out in its terms of appointment. The contract administrator is not a party to the contract, but is named in the contract (cl 11.5), and the extent of its authority to administer the contract derives from the wording of the contract.

3.5   The RIBA Building Contracts place various duties on the contract administrator, particularly with regard to issuing certificates or statements, as well as a wide variety of powers, such as the power to issue instructions. For a full list of these duties and powers, along with references as to where they are discussed in this Guide, see Tables 3.1 and 3.2.

3.6   In some matters the contract administrator will act as an agent of the client, for example when issuing instructions that will vary the works. In other instances it acts as an independent decision-maker, e.g. when deciding on claims for additional payment. When acting in the latter capacity, it would be implied that the contract administrator must act fairly at all times. It would be sensible for the contract administrator to use one of the RIBA standard forms of appointment to ensure that the obligations under its appointment align with those in the RIBA Building Contracts.

3.7   Failure by the contract administrator to comply with any obligation, either express or implied, may result in the contractor suffering losses. As the contract administrator is not a party to the contract, if the contractor wishes to bring a claim, this would, in the first instance, have to be against the client. It is likely, however, that any failure to administer the building contract according to its terms would be a breach of the contract administrator's duties to the client and, therefore, the client may seek, in turn, to recover its losses from the contract administrator.

Table 3.1 Contract administrator's duties

Clause CBC	DBC	Duty	Chapter reference
3.1	3.1	Attend the pre-start meeting	3.40
3.3.2	3.3.2	Take into account (when determining a revision of time and/or additional payment) any failure by the contractor to take reasonable steps to avoid the effects of an event	4.35, 4.37
3.6	–	Attend progress meetings (CBC only)	3.43
5.1	5.1	Administer the contract, issuing instructions and certificates and taking decisions	3.4–3.7, 5.10
5.2.1	5.2.1	Give the contractor two free copies of the contract documents	2.38
5.2.2	5.2.2	Issue any updates promptly	2.38
5.5	5.6	Issue instructions in writing	5.24
5.6.1	5.7.1	Confirm oral instructions to the contractor in writing promptly	5.24
5.13.1	5.10.1	Aim to agree any revision of time and/or additional payment promptly	4.29
5.13.2	5.10.2	Determine the appropriate adjustment to the contract time and/or price	4.29, 4.37, 6.13-6.15
5.14	5.11	Issue instructions to deal with inconsistencies in contract documents	2.44
7.3	7.1	Issue payment certificates, not later than 5 days after the due date (CBC), or at the frequency specified in the Contract Details (DBC)	6.24, 6.40, 6.58
7.11.2	7.8.2	Aim to agree the final contract price with the contractor within 90 days	6.54, 6.75
7.11.3	7.8.3	Issue a final payment certificate, if the parties cannot agree or the contractor makes no submission	6.54, 6.75
7.13.1	–	Issue a final payment certificate	6.56
9.11.1	9.11.1	Aim to agree on a revision of time promptly	4.29
9.11.2	9.11.2	Make a reasonable decision about a revision of time, if parties unable to agree or the contractor does not apply	4.28–4.31, 4.37, 4.42
9.12	9.12	Amend the date for completion and inform the parties	4.29
9.14	9.14	Aim to agree on an additional payment promptly	6.17
9.14	9.14	Make a reasonable decision about an additional payment, if unable to agree or the contractor does not apply; amend the contract price and inform the parties	6.17

Table 3.1 Contract administrator's duties – Continued

Clause		Duty	Chapter reference
CBC	DBC		
9.16	9.16	Certify practical completion or inform the contractor that it does not agree that it has been achieved	5.36–5.37
9.18	9.18	Issue a notice identifying areas to be taken over	5.41
10.3.2	10.3.2	Notify the parties when satisfied that defects are fixed	5.48
10.3.3	10.3.3	Issue the contractor with a notice if it fails to fix a defect	5.47–5.48
11.11.2	11.11.2	Issue instructions on communication procedures if the parties fail to set these out	3.42
A4.4	A4.1	When satisfied that a milestone has been achieved, include the payment for that milestone in the payment certificate for the relevant due date (CBC), or issue a payment certificate (DBC)	6.44–6.47, 6.66
A10.1	A8.1	Maintain a risks register	3.44

Table 3.2 Contract administrator's powers

Clause		Power	Chapter reference
CBC	DBC		
3.6	–	Invite people to the progress meetings	3.43
5.3.1	5.3.1	Visit the site, including any off-site locations in connection with the works	5.16
5.3.2	5.3.2	Inspect the works	5.16
5.3.3	5.3.3	Reject defective work	5.19
5.4	5.4	Issue instructions about a range of matters	2.44-2.45, 4.17, 5.13, 5.19-5.23
5.6.4	5.7.4	Instruct that oral instructions shall not take effect until confirmed in writing by the contract administrator	5.27
5.7	5.5	Instruct work is uncovered and inspected or tested	5.19, 6.11
5.8	5.5.3	Accept work not in accordance with the contract	5.20, 6.12
5.10	–	Modify, amend or withdraw an instruction, following a cl 5.9 notice from the contractor	5.29
5.11	5.8	Issue a 7-day notice to comply with an instruction	5.30
9.7	9.7	Instruct that work stops if force majeur occurs	–
12.1	12.3	Issue the contractor with a notice of intention to terminate	8.25

Role of the contractor

3.8   The contractor has the most extensive lists of duties and rights, which is unsurprising as it is the contractor which is primarily responsible for delivering the project on time and to the client's requirements. Full lists of these duties and rights are provided in Tables 3.3 and 3.4, and the key duties are outlined below.

Completing the works

3.9   The main obligation of the contractor is to complete the works as set out in the contract documents. In both CBC and DBC, this obligation is set out in clause 1.1, which states:

> The Contractor shall:
>
> 1.1.1 start the Works on the Start Date
>
> 1.1.2 carry out the Works regularly, diligently and in a good and workmanlike manner to ensure that they are completed properly in accordance with the Contract and all statutory requirements by the Date for Completion
>
> 1.1.3 be responsible for all statutory fees, notices and charges not covered in item K [CBC; item L, DBC] of the Contract Details.

3.10   The above clause does not refer expressly to 'materials, goods and workmanship', as would be usual in JCT contracts, and it is therefore suggested that 'in a good and workmanlike manner' refers to the method of carrying out the work, not what work is to be done. Therefore, it is essential that the standards for materials, goods and workmanship are set out clearly in the specification and other contract documents. Clause 1.2 states that:

> The Contractor shall use methods and products which minimise nuisance and pollution and are safe and fit for the purposes intended.

This is also intended to refer primarily to workmanship, but may also act as a useful catch-all to cover matters that have not been specified in detail. However, it is doubtful whether the obligation would override, for example, any express specification that had set out methods or materials that were unsafe.

Contractor's design obligation

3.11   Like most traditional contracts these days, the RIBA Building Contracts make provision for some design to be carried out by the contractor. It is an optional provision, but it is likely that on most projects it will be required to some extent. Any decision that affects the final form of the building (as opposed to the method of construction) is a design decision, even if it relates only to small levels of detail, such as the exact size of joist hangers or of central heating pipes. If in any doubt as to whether something is effectively 'design', the contract administrator should consider whether it is intending to make all the necessary decisions, and, if not, what would happen if the particular detail were to fail: would it be prepared to accept responsibility? If not, it should take steps to ensure that liability for design of that detail is expressly placed with the contractor.

3.12   The provisions are set out in optional clause A2, and the parts to be designed by the contractor are be described in item U (CBC) or item R (DBC) of the Contract Details. A 'Contractor's design proposal' is referred to in item F (both contracts), therefore the contract anticipates that the contractor may have provided a design prior to the contract being executed, which may be for some or all of the parts, depending on what was requested at tender stage (see paras. 2.15, 2.16 and 2.27). It appears that the contractor is responsible for designing only the parts that are listed. The contract makes it clear that the contractor is not liable for design provided by the client or the client's agents (A2.4).

However, it must notify the contract administrator of any discrepancy it finds in those designs.

3.13   What happens if the contractor makes a design decision relating to a part that has not been stipulated in the Contract Details, but for which the contract administrator has provided no design information? It is possible that the contractor would not be held responsible for that design decision. A relevant case concerns the Museum of Liverpool (*National Museums and Galleries on Merseyside* v *AEW Architects and Designers*), where the judge said the contractor was not responsible for designing anything not identified as part of the contractor's designed portion (CDP), in this case deciding the tolerance gaps between steps. Courts might a take less strict view in a smaller project, but to be safe it is best to use the 'Contractor Design' optional clause and to make sure the extent of the contractor's design responsibility is clear.

3.14   It is therefore very important that the delineation between the parts to be designed by the contractor and the rest of the project is described accurately.[1] This can be quite difficult in practice, especially where several parts or elements are listed, or the list includes a system (e.g. services) that is integral to many parts of the building.

3.15   The RIBA Building Contracts are clear that the contract administrator remains responsible for any integration. This could extend to the physical junctions between the contractor designed parts and other parts, but could also cover the combined performance of several systems or of systems and elements. If it is intended that the contractor is to be responsible for resolving any interface (physical or performance), the interface would have to be identified clearly within the parts to be designed by the contractor. The contractor retains the copyright in any design it provides, but grants the client a license to use it for the Works and related purposes (cl A2.5).

Contractor's design liability

3.16   Under clause A2 the level of liability of the contractor for design is to use the skill and care of a competent designer. The contractor is also required to ensure that its design meets the client's requirements. Taken together, this would not be interpreted as an absolute obligation to meet the specification, but a duty to use reasonable skill and care so that it does.

3.17   There are essentially two levels of liability used in construction. Professionals normally undertake to use reasonable skill and care, rather than promise to achieve a particular result. To understand this distinction, consider the case of a medical professional: a doctor would never promise to cure a patient, but simply to use their medical skills competently to achieve the best outcome possible. The fact that the patient does not get better is not of itself sufficient to show the doctor made a mistake. Similarly, with professional designers such as architects, the fact that there is a defect in the design is not sufficient to show they were negligent. Any client bringing a claim must also prove that the architect failed to use the reasonable skill and care of a competent architect. This is sometimes referred to as a 'negligence-based liability' and is the normal level of liability that would be set out in an architect's appointment or would be implied by the courts if no level has been set.

---

[1]   An example of the confusion that can arise when the contractor designed items are not accurately defined or described can be seen in *Walter Lilly & Company Limited* v *Giles MacKay and DMW Ltd.*

3.18   The alternative, more onerous, level of liability is to promise to achieve a result. This is often referred to as a 'strict liability', as there is no need for the claimant to show that there was any negligence on the part of the designer; it would be enough to show that the result did not meet the stated requirements in some respect. A commonly used shorthand descriptor is that the designer has taken on a 'fit for purpose' liability. Contractors who undertake design work often take on an express 'fit for purpose' liability. Where no level of liability had been set out, the courts may imply that there is a strict liability, provided it can be shown the client was relying on the contractor's skill and judgment.

3.19   The RIBA Building Contracts contain an option to set a strict liability obligation, in place of reasonable skill and care. This appears in the Contract Details under 'Contractor Design' (item U in CBC, or item R in DBC), which states 'Contractor's design to be fit for purpose'. No default is given: the parties are required to indicate 'yes' or 'no'. However, the effect of clause A2.3 is that the default would be 'no', as it states:

> If the option 'Contractor's design to be fit for purpose' is selected … then clause A2.1.2 shall not apply and the Contractor's design shall be fit for the purpose indicated in the Contract Documents.

3.20   Clients would usually prefer a 'fit for purpose' level of liability. After all, if they have taken the trouble, with their consultant, to set out detailed requirements for the design, including specific performance targets (e.g. in relation to energy use), they are likely to want to be able to bring a claim should the building not perform, without having to also prove a lack of care. However, as the 'fit for purpose' level of liability is more onerous on the contractor, some contractors may be unwilling to tender on this basis or might submit higher tender prices. The JCT contracts do not generally contain this option (except in the case of the Major Project Construction and Constructing Excellence contracts). However, it is an optional provision in NEC3 contracts (and standard in the FIDIC and IChemE contracts), and is often asked for in bespoke contracts, therefore its use is increasing.

3.21   Under clause A2.7 the contractor is required to 'ensure there is adequate professional indemnity insurance for its design responsibilities, as set out in … the Contract Details' (if no requirement is set out the obligation would fall away). The contractor might not be able to obtain insurance to cover a 'fit for purpose' risk, but that would not, of course, affect its liability.

3.22   Note that the contractor is also required to 'compensate the [client] for all claims in respect of the Contractor's designs' (cl A2.6). This is a wide liability – it would cover, for example, claims by tenants for repair work or claims for breach of copyright – wider than a normal professional liability for design errors. There is no requirement to insure to cover this liability, and if the contractor's design is extensive the client should consider whether to add this requirement at tender stage.

Contractor's obligations in respect of subcontracted work

3.23   The contractor's duties with respect to subcontracting the work are set out briefly in the following two clauses:

> 1.6 The Contractor shall inform the Architect/Contract Administrator of any parts of the Works that it has subcontracted.

> 1.7 The Contractor is solely responsible for carrying out the Works and for the performance of all subcontractors and suppliers.

3.24 The contractor may subcontract to anyone it chooses; there is no requirement to inform the contract administrator beforehand, or to obtain permission from the contract administrator or the client, and neither has the power to bar a contractor from using a particular firm. The contractor is, of course, entirely responsible for all its subcontractors (cl 1.7), so in theory there is no risk to the client, even if the contractor selects firms that are not capable of achieving the desired standard of work. However, in practice it can be frustrating to stand by without intervening, as work has inevitably to be redone and delays are caused.

3.25 If the client wishes to take a more proactive stance, there are two possible courses: one is to add a clause requiring the client's or contract administrator's permission for any sub-contracting (or perhaps only for subcontracting certain work), with such permission not to be unreasonably withheld. The second is to use the 'Required Specialists' provisions for critical areas of work.

Required specialists

3.26 Under an optional clause (A7 in CBC and A5 in DBC) the client may specify that the contractor uses particular specialists to carry out described parts of the works in the contracts. If used, the contractor is fully responsible for the performance of those specialists (cl 1.7 and A7.1.3, CBC; cl 1.7 and A5.1.3, DBC), in the same way that it would be for its own domestic subcontractors.

3.27 The details of the specialists, and the work they are to carry out, are entered in the Contract Details (item Z in CBC, or item U in DBC). Clause A7.1.1 (CBC) and clause A5.1.1 (DBC) require that the contractor should be given 'reasonable advance notice of the Required Specialists and is deemed to be satisfied with their suitability'. What is 'reasonable notice' will depend on the circumstances. Normally a reasonable tender period should be sufficient; however, if possible, it may be prudent to alert contractors at an earlier date, i.e. before the formal tender invitations are sent out.

3.28 The notice period is included to allow the contractor to make enquiries as to the suitability of the firm, and its price and terms for carrying out the work. The requirement to enter into a subcontract with the required specialist (cl A7.1.2, CBC; cl A5.1.2, DBC) is said to be 'subject to clause A7.1.1 [CBC; A5.1.1, DBC]'; therefore, if inadequate notice is given (e.g. the requirement is a late addition to the tender information or is introduced after the tenders have been returned), the contractor can refuse to engage the specified firm.

3.29 It is also possible that if the contractor is not given adequate notice, it may engage the specialist, but then attempt to claim for additional time and money should the firm later default on its subcontract (there is old case law that may support such a claim, for example see *Gloucestershire County Council* v *Richardson*). It is suggested that in the context of the RIBA Building Contracts the argument is unlikely to succeed: clause 1.7 and clause A7.1.3 (CBC) or A5.1.3 (DBC) are unqualified, and it is the requirement to engage the specialist, not the responsibility once engaged, that is subject to reasonable notice.

Gloucestershire County Council v *Richardson* [1969] 1 AC 480 HL

In this case the House of Lords considered whether a main contractor might be liable to the employer in respect of latent defects in materials delivered by a nominated supplier. It held that the main contractor's liability to the employer was limited to the extent of the nominated supplier's liability to the main contractor by operation of the terms of the nominated subcontract.

3.30   If the contractor terminates the required specialist's employment, it must notify the contract administrator, who is required to 'issue appropriate instructions regarding a replacement'. The contract then states that the contractor 'shall be entirely responsible for any delay or additional costs arising from the termination'.

3.31   This juxtaposition is not particularly comfortable. If the contractor is responsible for the specialist firm, and the losses, then it would be more straightforward if it was given the responsibility of finding a replacement, subject to the contract administrator's approval, and confirming instruction. However, there is nothing to prevent the contractor from making proposals, which would be sensible, as they may minimise the losses it will suffer. Nevertheless, the contract administrator is responsible for taking action and resolving the situation.

3.32   The contracts do not distinguish between the possible different reasons for the termination, so it appears that the liability would apply whatever the reason; whether a simple falling out or a disagreement about money, or because the firm seriously failed to perform. Furthermore, they do not discuss what would happen if the required specialist refused to undertake the project, or became insolvent, and yet no termination was issued. The contractor is only made liable for consequences of any termination, not of complete non-performance, and a contractor might argue that in these circumstances the client should carry the risk. It would therefore be sensible for the contract administrator to take action promptly, whatever the cause of the termination.

Compliance with statute/health and safety legislation

3.33   The contractor is required to comply with all statutory requirements (cl 1.1.2). In addition, clause 4.3 states that 'the Parties shall comply with all health and safety regulations, including making any appointments or submissions required under such regulations'. CBC also requires the contractor to notify the contract administrator if, in its opinion, an instruction may have adverse health and safety implications (cl 5.9.2). It is not clear why this useful provision was not also included in DBC, nevertheless it is suggested that the contractor would normally be under a duty to warn of any instruction that would have an adverse effect.

3.34   Clause 1.1.3 states the contractor shall: 'be responsible for all statutory fees, notices and charges not covered in item L [CBC; item K, DBC] of the Contract Details'. However, it is important to note that it is not just the *statutory fees, notices and charges* that are referred to in item L or K, as this states: 'Responsibility for obtaining and paying for all planning permissions, Building Regulations approvals and party wall consents shall be taken by: [select Employer/Customer (default) or Contractor]'.

Clause		Duty	Chapter reference
CBC	**DBC**		
1.1.1	1.1.1	Start the works on the start date	3.9, 4.4
1.1.2	1.1.2	Carry out the works regularly, diligently and in a good and workmanlike manner	3.9, 4.4, 4.16, 4.20
1.1.2	1.1.2	Complete the works according to the contract and all statutory requirements	3.9, 3.33
1.1.2	1.1.2	Complete the works by the date for completion	3.9, 4.16
1.1.3	1.1.3	Pay all statutory charges etc. unless the contract documents state otherwise, and possibly obtain planning permission, Building Regulations approval and party wall consents	3.34
1.2	1.2	Use methods and products that avoid pollution, are safe and are fit for purpose	3.10
1.3	1.3	Ensure a suitably qualified representative is on site	5.4
1.4	1.4	Liaise with the employer/customer to maintain security	4.2
1.6	1.6	Inform the contract administrator of any parts of the works it has subcontracted	3.23
3.1	3.1	Attend the pre-start meeting	3.40
3.2	3.2	Provide the employer/customer and the contract administrator with a warning of any event that will affect progress of the works, and work together with the employer/customer to resolve the event	4.22
3.3	3.3	Take reasonable steps to minimise the effect of the event	4.23
3.6	–	Attend progress meetings	3.43
4.2	4.2	Obtain written consent of the employer/customer before assigning burdens, rights or benefits	–
4.3	4.3	Comply with all health and safety regulations	3.33, 5.6–5.8
5.5	5.6	Comply with instructions immediately	5.24
5.6.2	5.7.2	Issue a written record of an oral instruction to the contract administrator	5.24, 5.27
5.10	–	Comply with contract administrator's decision under clause 5.9	5.29
5.11.2	5.8.2	Co-operate with others engaged by the employer/customer following contractor failure to comply with a 7-day notice	5.30
5.11.3	5.8.3	Allow others (as detailed above) access to the site	5.30
5.12	5.9	Calculate the effect of a change to works instruction on contract price and completion date, and submit within 10 days	6.13
5.13.1	5.10.1	Aim to agree any revision to time or contract price promptly	4.29
5.14	5.11	Inform the contract administrator immediately of any inconsistencies it finds in the contract documents	2.44

Table 3.3 Contractor's duties

Table 3.3 Contractor's duties – Continued

Clause		Duty	Chapter reference
CBC	DBC		
6.3	6.3	Maintain insurance in respect of its liabilities to the value specified, policies to be in joint names	7.15–7.16, 7.17, 7.21
6.4	6.4	Provide evidence of insurance policies	6.36
–	7.4	Submit an application for payment if the contract administrator fails to issue a certificate (expressed as a right in CBC (cl 7.2/7.6.1))	6.81
7.8.1	7.7.1	Issue the employer/customer with a valid VAT invoice	6.49, 6.67
7.11.1	7.8.1	Submit its calculation of the final contract price, along with supporting documentation, not later than 90 days after completion	6.53, 6.74
7.11.2	7.8.2	Aim to agree the final contract price with the contract administrator within 90 days	6.54, 6.75
7.15	7.9.1	Pay any amount indicated on the final payment certificate or application	6.56, 6.83
9.7	9.7	Inform the client and the contract administrator if force majeure occurs	–
9.10.1	9.10.1	Apply for any revision of time within 10 days of an event occurring	4.27
9.10.2	9.10.2	Apply for any revision of time within 10 days of the end of a continuing event	4.27
9.11.1	9.11.1	Aim to agree on a revision of time promptly	4.29
9.13.1	9.13.1	Apply for any adjustment to the contract price within 10 days of an event occurring	6.16
9.13.2	9.13.2	Apply for any adjustment to the contract price within 10 days of a continuing event	6.16
9.14	9.14	Aim to agree on an additional payment promptly	6.17
9.16	9.16	Notify the contract administrator when it thinks that practical completion of the works or a section has been achieved	5.36
10.3.1	10.3.1	Remedy all defects identified during defects fixing period	5.38, 5.46
11.11.1	11.11.1	Set out communication procedures at the pre-start meeting	3.40
13.1	13.1	Meet and negotiate to resolve any disagreements	9.2
–	13.2	If mediation is selected, refer disputes first to the mediator (a right under CBC (cl 13.3))	9.7
–	13.3	If adjudication is selected, refer disputes to the adjudicator (a right under CBC (cl 13.2))	9.10
A1.1	A1.1	Submit a programme to the contract administrator, at least 21 days before the start date	4.8–4.10
A1.2	–	Submit an updated programme to the contract administrator, at least 5 days before each progress meeting	4.18
A1.3	–	If further changes are agreed at the progress meeting, issue the updated programme promptly	4.18

Table 3.3 Contractor's duties – Continued

Clause CBC	DBC	Duty	Chapter reference
A2.1.1	A2.1.1	Design the parts of the works described in item U/item R; integrate the design with the works	2.26, 3.12
A2.1.2	A2.1.2	Use reasonable skill, care and diligence in the design	3.16
A2.1.3	A2.1.3	Ensure the design is in accordance with the employer/customer's specification	3.16, 5.14
A2.2	A2.2	Submit details of its design to the contract administrator	5.12
A2.3	A2.3	Ensure the design is fit for purpose	3.19
A2.4	A2.4	Notify the contract administrator of any discrepancies it finds in the employer/customer's or architect's designs	3.12
A2.6	A2.6	Compensate the employer/customer for all claims in respect of its designs	3.22, 7.12
A2.7	A2.7	Ensure there is adequate professional indemnity insurance for its design	3.21, 7.24
A5.2	–	Provide an advanced payment security	6.20
A7.1.2	A5.1.2	Employ the required specialists to carry out the described parts of the works	3.26
A7.1.4	A5.1.4	Notify the contract administrator of the termination of the employment of a required specialist	3.30
A8.1	–	Execute a collateral warranty or third party rights agreement in favour of identified parties	2.29
A9.1	–	Adhere to all legislation relating to official secrets	–
A9.2.1	–	Pass all requests for information under the Freedom of Information Act 2000 to the employer	–
A9.3.1	–	Ensure that discrimination in any form is not practised or allowed	–
A9.3.2	–	Ensure that no form of bribery and/or corruption is engaged in	–
A11.1	A9.1	Set out the rules that the contract administrator must follow when assessing a revision of time or an additional payment	2.31

Table 3.4 Contractor's rights

Clause CBC	DBC	Right	Chapter reference
1.5	1.5	Use, free of charge, any employer/customer's facilities listed in the contract documents	2.13
3.4	3.4	Propose changes to the works that may improve their value or lower the contract price	6.9
5.9.1	–	Notify the contract administrator that an instruction is not in accordance with the contract	5.29

Table 3.4 Contractor's rights – Continued

Clause		Right	Chapter reference
CBC	DBC		
5.9.2	–	Notify the contract administrator that an instruction would have an adverse effect on health and safety	5.29
5.9.3	–	Notify the contract administrator that an instruction would have an adverse effect on any part designed by the contractor	5.29
7.2	–	Issue applications for interim payments (this is expressed as a duty under CBC (cl 7.2), where the contract administrator fails to issue a certificate)	6.22, 6.65
7.6.1	–	Issue a payment notice in the absence of a payment certificate	6.80
7.13.2	7.8.4	Issue a final payment notice or application in the absence of a final payment certificate	6.80, 6.83
8.1.1	8.1.1	Issue a notice of its intention to suspend some or all of its obligations	6.87
8.1.3	8.1.3	Suspend some or all of its obligations	6.86
9.9	9.9	Apply for a revision of time	4.25–4.33
9.13	9.13	Apply for an adjustment to the contract price	6.16
12.3	12.5	Issue a 14-day notice of intention to terminate	6.89, 8.16
12.4	12.6	Terminate its employment by issuing a notice of termination	6.89, 8.16
12.5	12.7	Terminate its employment by issuing a notice of termination	8.20
13.2	–	Refer disputes to adjudication (this is expressed as a duty under DBC (cl 13.3))	9.10
13.3	–	If mediation is selected, refer disputes to mediation (this is expressed as a duty under DBC (cl 13.2))	9.7
13.4	13.4	If arbitration is selected, refer disputes to arbitration	9.21
13.5	13.5	If arbitration is not selected, refer disputes to the appropriate court	–

3.35   This is, of course, a far greater responsibility than is set out in clause 1.1.3. If the client wishes the contractor to obtain permissions, it would be important to make this clear in the tender documents, and to ensure that 'Contractor' is selected in item L (CBC) or item K (DBC), as otherwise this would not be covered by the lesser obligation in clause 1.1.3.

Role of the client

3.36   The client has a significant role in any construction project. Its key duties, outlined in Table 3.5, are ones of collaboration. For example, not to take any action that would interfere with the carrying out of the works, and to pay the contractor on time in accordance with the contract. The client is also given various rights, as listed in Table 3.6. Generally, though, the client should discuss both its duties and its rights with the contract administrator, and should seek the contract administrator's advice before taking any action; the contract administrator has the overall responsibility for administering the project, and if actions are co-ordinated there is less likelihood of clashes or unexpected consequences.

Table 3.5 Client's duties

CBC	DBC	Duty	Chapter reference
2.1	2.1	Allow contractor reasonable access to the site for pre-construction inspection, carrying out the works and rectifying defects	4.3, 5.38
3.1	3.1	Attend the pre-start meeting	3.40
3.2	3.2	Provide the contractor and the contract administrator with a warning of an event that will affect progress of the works, and work together with the contractor to resolve the event	4.22
4.3	4.3	Comply with health and safety regulations	5.6–5.8
6.3.1	6.3.1	Maintain insurance in respect of its liabilities to the value specified	7.15, 7.16, 7.20
6.4	–	Take out insurance if the contractor fails to provide evidence, and deduct the cost from amounts due to contractor (expressed as a right in DBC)	6.36
7.7	–	Pay amount due on any payment certificate or payment notice on the final date for payment	6.48, 6.49, 6.80
7.8.2	7.3.2	Pay the contractor's invoice promptly	6.49, 6.67–6.69
7.15	7.9.1	Pay any amount indicated on the final payment certificate	6.56, 6.83
9.7	9.7	Inform the contractor and the contract administrator if force majeure occurs	–
11.11.1	11.11.1	Set out communication procedures at the pre-start meeting	3.42
13.1	13.1	Meet and negotiate to resolve any disagreements	9.2
–	13.2	If mediation is selected, refer disputes to the mediator (expressed as a right in CBC (cl 13.3))	9.7
–	13.3	If adjudication is selected, refer disputes to adjudicator (expressed as a right in CBC (cl 13.2))	9.10
A4.1	–	Pay the contractor the contract price when the contract administrator has certified practical completion	6.39
A5.1	–	Pay the contractor an advance payment	6.20
A6.1	–	Provide evidence of its ability to pay the contract price	–
A11.1	A9.1	Set out the rules that the contract administrator must follow when assessing a revision of time or an additional payment	2.31

Table 3.6 Client's rights

CBC	DBC	Right	Chapter reference
2.2	2.2	Replace the appointed contract administrator	2.14
2.3	2.3	Appoint others under item E	2.14
2.4	2.4	Defer access to the site or sections of the site	4.6

Table 3.6	Client's rights – Continued		
3.4.1	3.4.1	Seek the contract administrator's advice on any changes proposed by the contractor	6.9
3.4.2	3.4.2	Accept the proposed changes	6.9
5.11.1	5.8.1	Employ others if the contractor fails to comply with a 7-day notice	5.30, 6.36
–	6.4	Take out insurance if the contractor fails to provide evidence, and deduct the cost from amounts due to the contractor (expressed as a duty in CBC)	6.36
7.9.1	–	Issue a pay less notice	6.51
7.10	–	Issue a pay less notice; authorise another to issue a notice on its behalf	6.51
9.17	9.17	Request to use part of the works for storage or other purposes	5.40
9.18.1	9.18.1	Request to take over part of the works before practical completion	5.41
10.1	10.1	Deduct liquidated damages	5.45, 6.36
10.3.4	10.3.4	Employ others if the contractor fails to rectify a notified defect	5.47, 6.36
–	12.1	Cancel the contract within 14 days	1.27
12.1	12.3	Issue a 14-day notice of intention to terminate the contractor's employment	8.6
12.2	12.4	End the contractor's employment with a termination notice	8.6
12.5	12.7	End the contractor's employment with a termination notice	8.20
13.2	–	Refer disputes to adjudication (expressed as a duty in DBC (cl 13.3))	9.10
13.3	–	If mediation is selected, refer disputes to mediation (expressed as a duty in DBC (cl 13.2))	9.7
13.4	13.4	If arbitration is selected, refer disputes to arbitration	9.21
13.5	13.5	If arbitration is not selected, refer disputes to the appropriate court	9.21
A1.4	A1.2	If financial consequences have been selected, withhold a percentage of any amount due until the contractor submits a programme	4.10, 4.19
–	A6.1	If item V is selected, to act as the contract administrator	3.37

Customer acting as contract administrator

3.37   DBC is unusual in that it offers the facility for the customer to name themself as contract administrator (optional cl A6). Although it is sometimes frowned upon, there is no reason why the customer can't name themself in this way, with the very important proviso that it must have been made completely clear to the contractor at tender stage that this was going to be the arrangement. The customer must also ask themself whether they have the skills and resources (particularly time) necessary to perform the role, and whether they are capable of being truly impartial. Any incorrect or biased decisions are liable to be challenged by the contractor and, if a reasonable solution is not agreed, may result in the contractor being successful in subsequent litigation (as the tribunal is likely to examine the decision very closely). It is unwise for a customer to hope that it may secure a 'better deal' by taking on the role of contract administrator.

Management systems

3.38   The RIBA Building Contracts place more emphasis on management, and have more provisions concerning meetings, co-ordination and communication, than do other equivalent contracts.

3.39   These provisions are aimed at ensuring the smooth running of the project. They include requirements to:

- provide and update a programme;
- give early warning of matters that may cause a delay or affect the contract price (see paras. 4.22–4.23);
- suggest improvements (see para. 6.9);
- hold a pre-start meeting;
- hold progress meetings;
- maintain a risks register.

Some of these mechanisms are discussed in other parts of this Guide, but the last three are of more general application.

Pre-start meeting

3.40   The requirement to hold a pre-start meeting, at least ten days before the work is due to start on site, is contained in clause 3.1. Both parties must attend, as well as the contract administrator. Clause 3.1 states that the parties must:

3.1.1   set out expectations for each Party

3.1.2   set out communication procedures, including any specific rules on electronic communications

3.1.3   identify risks and set out the risk mitigation procedure

3.1.4   set out any administrative procedures.

3.41   The pre-start meeting is often the point at which the contract documents are signed. None of the discussions ought to result in any changes to the contract terms or the obligations of the parties. However, it would be wise to bear in mind that there is nevertheless a close link between the above and various contractual issues. For example, the 'expectations' cannot be more than those that are already set out. The client, for example, cannot demand a higher standard of work than the level specified in the contract documents or that the contractor finishes earlier than the contractual date. If it becomes apparent at the meeting that expectations are at odds with those in the contract, then the documents should be amended before signature.

3.42   The meeting will be used to flesh out and fine-tune some of the procedural matters. Key among these would be whether the client is to remain in occupation, and what the exact arrangements will be with regard to security, safety, access, use of facilities, storage and disposal of rubbish. If any additional or new arrangements are made, these should be recorded and annexed to the contract documents. The same applies to communications and other administrative procedures; the meeting will also be used to agree communication procedures (cl 11.11.1); and if the parties fail to do this the contract administrator is required to issue instructions regarding these (cl 11.11.2). Generally all communication is to be in writing and issued to the parties and the contract administrator (cl 11.10), unless the contract states otherwise. Special procedures are set out in the case of termination (cl 11.12, see para. 8.26).

Progress meetings (CBC)

3.43   CBC requires that progress meetings are held monthly (or other period as agreed) and attended by the contractor, the contract administrator and any other party invited by the contract administrator (cl 3.6). The meetings should cover all operational issues relating to the project. It is common practice to hold such meetings, but including them in the contract means that it becomes an obligation on the contractor to attend, and a breach if it does not. The meeting will normally cover matters such as progress, technical issues and information needed, and is an opportunity to discuss advance warning notices. In particular, the contractor is required to provide an update to its programme five days before the meeting, which is a useful way for the administrator to monitor and record how the job is progressing. Parties using DBC may wish to consider adding in this provision.

Risks register

3.44   This is an optional provision that, if selected, requires the contract administrator to establish and maintain a risks register (cl A10 in CBC, or A8 in DBC). Initially, the register will list the risks and mitigation procedures identified and agreed at the pre-start meeting, but these are likely to be adjusted as the project progresses and new risks are identified. The register could be in the form of a simple list (or more likely a spreadsheet), and it is common practice to place a priority on the risks (e.g. highly likely, not likely), as well as to set out the measures or actions to be taken should a risk materialise and who will take them. Strictly speaking there is no contractual obligation to comply with any actions set out (unless, of course, the register includes obligations already covered in other contractual provisions). Nevertheless, it may be a useful point of reference in day-to-day communications and a good discussion item at progress and other meetings.

4 Project progress

4.1 Most building contracts include provisions that require the contractor to complete by a specific date or set of dates as this is often a matter of great importance to the client. Late completion may well result in losses to the client, in particular the costs of alternative accommodation and additional consultants' fees. However, the obligation to complete is normally subject to some exceptions; for example, the client will be required to allow the contractor additional time if the client itself causes delay. This chapter examines the contractor's obligations regarding time, programming and completion under the RIBA Building Contracts, including the mechanisms for monitoring progress and the sanctions for non-completion.

The site: possession

4.2 An address for the site is to be given in item C of the Contract Details. However, far more information may be needed than simply the address. If the contractor is not to have full access to the whole plot at that address – for example, if only part of the plot can be used, or if there are restrictions on entry points – this should be made clear at the tender stage. Similarly, if the building is to be occupied, the details of this should be given, for example which parts and between which dates. Where the property shares common parts with other properties (e.g. stairs, parking), it may be sensible to explain exactly what the contractor may use of these. The contractor is required to liaise with the client regarding security (cl 1.4) – it would be sensible to set out as much information as possible on responsibility for security in the Contract Details.

4.3 The client has to allow 'reasonable access' to the contractor for carrying out the works (cl 2.1.2). Unlike some of the JCT contracts, the RIBA Building Contracts do not refer to the contractor having 'possession' of the site (which is normally held to be a licence to occupy the site up to the date of completion; *H.W. Nevill (Sunblest) Ltd* v *William Press & Son Ltd* and *Impresa Castelli SpA* v *Cola Holdings Ltd*). In the RIBA Building Contracts, 'reasonable access' will be interpreted in the light of all the information made available to the contractor at the time of tender. If little information is given, the courts will imply an obligation that the contractor should be given such possession, occupation or use as is necessary to enable it to perform the contract (*London Borough of Hounslow* v *Twickenham Gardens Development*). This may include access not just to the building where the work is to be carried out, but also to other areas in the control of the client (see *The Queen in Rights of Canada* v *Walter Cabbott Construction Ltd*).

The Queen in Rights of Canada v *Walter Cabbott Construction Ltd* (1975) 21 BLR 42

This Canadian case (Federal Court of Appeal) concerned work to construct a hatchery on a site (contract 1), where several other projects relating to ponds were also planned (contracts 3 and 4). The work to the ponds could not be undertaken without occupying part of the hatchery site.

> Work to the ponds was started in advance of contract 1, causing access problems to the contractor when contract 1 began. The court confirmed (at page 52) the trial judge's view that 'the "site for the work" must, in the case of a completely new structure comprise not only the ground actually to be occupied by the completed structure but so much area around it as is within the control of the owner and is reasonably necessary for carrying out the work efficiently'.

Starting the work

4.4   The contractor is required to start on the start date entered in the Contract Details (cl 1.1.1; note: there may be several start dates if optional clause A3 is adopted). The contractor must then carry out the works regularly and diligently (cl 1.1.2). Note that this is similar to the approach in the SBC and IC11 contracts, but different to that in the MW11 version; in the latter the contract simply gives a date when 'work may be commenced'.

4.5   The requirement to start on a particular date is useful, as it is often crucial that work actually starts on the defined date. In cases where the site is empty, the client needs to be assured that from that date the contractor will be responsible for security, health and safety, and general compliance with local authority requirements. An empty site is exposed and hazardous, so the client will be at risk of claims if security is not being addressed. Even where a customer intends to remain in residence, it is disconcerting, and sometimes extremely inconvenient, if work does not commence on the planned date.

4.6   There are provisions to allow the client to defer access to the whole site, or to sections of the site, if needed (cl 2.4). This can be very helpful if, for example, the client has problems arranging alternative accommodation, or for removal of furniture or equipment to storage. The contract does not place any limits on the length of deferral allowed, nor does it require the client to give any advance warning notice. However, the contractor would be entitled to a revision of time, and to additional payment to cover costs arising as a result of the delay. As these could be considerable, especially if little or no warning is given, the client would be wise to exercise the right to defer only in emergency situations, or if the deferral and the contractual implications can be agreed with the contractor well in advance.

Completion in sections

4.7   The RIBA Building Contracts also include an optional clause (cl A3) providing for completion in sections. Although the provisions refer only to 'Completion', it is possible to arrange for the work to be started and/or finished in sections. This would be useful where the client cannot make all areas of the site available at the same time, or needs certain parts of the building before others. On large projects, phased working is quite often more economical for the contractor, as it can move its resources around the various sections in a phased programme. On smaller projects, however, it may actually be less convenient, so this may need to be negotiated following tender submissions. If this option is selected, separate start and completion dates and rates of liquidated damages are specified for each section in item V (CBC) or item S (DBC) of the Contract Details.

The contractor's programme

4.8   Both versions of the contract contain an optional clause (cl A1), whereby the contractor is required to provide a programme. It is recommended that this should always be selected (optional Item T CBC and Q DBC), except on the very smallest of projects. The programme is of great help for:

- giving the client a general idea as to what to expect (especially important to the client if they are in residence);
- giving the contract administrator an indication as to when the contractor will require further information;
- alerting the contract administrator as to when it may be wise to inspect the site;
- acting as an early alert if the contractor is slipping behind programme or getting into financial difficulties.

4.9   The programme is not identified as being one of the contract documents under item F of the Contract Details. Nevertheless, it would be sensible to ask that the contractor provides a programme before the contract is entered into. This could be done either by requiring tenderers to provide a programme with their tenders, or, once the tenders have been received, by requesting the preferred bidder to submit a programme before their tender is accepted. Otherwise, the programme is to be provided 21 days before the start date (cl A1.1 CBC and DBC).

4.10   There are alternative sanctions available for non-production of a programme (a selection is made in the Contract Details, item T CBC or item Q DBC). The contractor can either be subject to a financial penalty or prevented from starting work until the programme is produced. The contractor would be still be bound to finish by the completion date, despite the delayed start. The financial penalty involves withholding 10 per cent of the value of the first payment certificate until the programme is produced (cl A1.4 and A1.5 CBC, A1.2 and A1.3 DBC). The money is, of course, paid when the programme is produced (presumably immediately, although the contracts do not say). This will have a significant impact on the contractor's cash flow, but ultimately the financial penalty will only be the interest on the withheld amount. The interest on 10 per cent for a few weeks will not be a large sum, whereas the losses that could arise due to being unable to start are likely to be much higher (i.e. the liquidated damages for that number of weeks, plus possible cancellation and start-up costs), therefore this is likely to be a more effective sanction. It is notable that NEC3 has a higher deduction: 25 per cent of any amount due (cl 50.3).

Content of programme

4.11   The parties are required to set out in the Contract Details, under item T in CBC or item Q in DBC, what the programme should contain (this should be done at tender stage). Two options are listed:

The activities the Contractor will carry out to complete the Works, the start and finish dates of each activity and the relationship of each activity to the others, including lead and lag times.

The number of people and other resources for each activity.

4.12   It is possible to select both options, and to add further requirements. The first will, among other things, give a clear idea of the sequence of work operations throughout the project. Including 'the relationship of each activity to the others' will show whether an item needs to be finished before another can start, i.e. which items are time critical. The shortest route through all time critical activities is usually referred to as the 'critical path'. Knowing the critical path is extremely useful when it comes to assessing revisions of time. It should be noted that the critical path is not fixed and may change throughout the project, therefore having regular programme updates is essential.

4.13   The number of people and other resources is also very useful to know. It will give an immediate indicator of whether the contractor is not resourcing the project as planned, and therefore may be evidence that the contractor is responsible for a delay. In extreme cases it may signal that the contractor is getting into financial difficulty.

4.14   It is difficult to see why both options would not always be required. These days most contractors use software packages to work out their programmes, and the packages would always show both sets of information (in fact, an input of resources is required in order for the package to calculate the durations of activities).

Drawings/information required/provided

4.15   One of the further requirements for the programme might be to show the dates when the contractor will require additional information or drawings. There is no requirement in the contracts for further information to be provided, but it is likely that such a duty would be implied (see para. 5.10). It would therefore be useful for the contract administrator to be aware of when the contractor anticipates it will need information. Nevertheless, as the programme is not a contract document, any dates shown would not be binding.

Progress

4.16   The contractor is required to proceed regularly and diligently and to complete by the date for completion (cl 1.1.2). The meaning of this phrase is discussed in detail in Chapter 8, but essentially it means maintaining steady progress using adequate resources. The contractor does not have to stick precisely to its own programme, so long as it completes by the date for completion; if the client requires any parts to be completed before others, or to use any parts during the course of the works, it will need to make use of the sectional completion provisions, or those for early use and partial possession before practical completion (see paras. 5.40–5.44).

4.17   The contract administrator will not normally intervene in matters concerning the day-to-day programming. As the contract administrator is given wide powers to issue instructions on any matter (cl 5.4.8) and the specific power to issue 'instructions on postponing the Works or Sections of the Works' (cl 5.4.4) it would be entitled to alter the working sequence, but should generally only use these powers where no other option is open. Examples might be if the local authority requires work to cease for a period of time, if an unanticipated health and safety hazard is encountered, or if the client has second thoughts about the design of a part of the project and work has to be put on hold while discussions are held. There will almost always be significant effects on the completion date and the costs of the project.

Updated programmes

4.18   CBC, but not DBC, requires the contractor to submit updated versions of the programme (the requirement is linked to progress meetings; there are no such meetings in DBC). In

CBC, the contractor is required to submit an updated programme to the contract administrator no later than five days before each progress meeting, and a revised version shortly after if amendments are agreed (cl A1.2 and A1.3). This is the system that is used in the Government General Conditions of Contract for Building and Civil Engineering (GC Works) contracts, and which has proved to be very effective in practice (it is also used in NEC3). For clarity, the 'update' should be provided even if there are no changes to the programme. It is very unlikely that this will be the case, however. Even without any delays being experienced, the contractor will in practice regularly review and make adjustments to the resourcing and sequencing, and it is very useful for the contract administrator to be made aware of these. It may be sensible to add a similar requirement to DBC, possibly for a monthly update to be provided, as a programme can quickly become out of date.

4.19   There is a financial penalty for late provision of updates, in addition to that for late initial production of the programme; for late updates, five per cent is withheld from certified amounts (cl A1.4 CBC). In practice, it is suggested that if no programme is produced for the first three months, this would mean withholding 10 per cent of the first certificate amount, and then five per cent of the second and the third. A financial penalty for late updates is also included in DBC (A1.2), even though there is no requirement to provide updates, i.e. there is no equivalent to CBC clauses A1.2 and A1.3. The parties could, of course, agree to add such a provision if they wish.

Finishing the work

4.20   The Contract Details require the insertion of a 'Date for Completion' (and if completion in sections is selected, a date for completion for each section), and the contractor is required to complete by this date (cl 1.1.2). However, this is rarely the date the works are actually completed. If delays occur that were not the fault of the contractor, the contractor will be entitled to a 'Revision of Time'; this will mean that the date for completion of the works or a section will be postponed, and the contractor's obligation will be to meet the revised date. If the contractor simply fails to finish by the original or revised date for completion, liquidated damages will be claimable by the client (note: the damages may need to be repaid if the date for completion is subsequently revised).

4.21   Once the works are complete, the contract administrator will certify 'Practical Completion', which is then followed by a 'Defects Fixing Period'. When all defects have been corrected, the contract administrator will issue a final payment certificate. The rest of this chapter considers delays to the programme. Practical completion and the remedying of defects is dealt with in Chapter 5.

Delay

4.22   Delays to a project will obviously cause problems for all involved, including the client, the contractor and the contract administrator, and so should be prevented or minimised if at all possible. The 'Collaborative Working' section that appears in both of the RIBA Building Contracts includes provisions that are aimed at preventing or managing delay and its consequences, under the heading 'Advance Warning and Joint Resolution of Delay'. The relevant clauses state:

    3.2 If an event affects or will affect progress of the Works and/or the Contract Price, the Parties shall:

3.2.1 provide each other and the Architect/Contract Administrator with an advance warning notice of the event as soon as they become aware of it

3.2.2 work together to resolve the event. If necessary the Architect/Contract Administrator shall hold a meeting with the Parties and other related stakeholders to resolve the event.

4.23   The obligation applies to both parties; so, for example, the client might notify of a possible restriction to access, and the contractor might notify of a problem with obtaining a material. A notification by the contractor must be given whether or not the event might entitle it to a revision of time. There is no obligation on the contract administrator to issue a notification of events of which only it is aware, but of course it would be sensible for it to do so. Also, although the clause states 'affects or will affect', it would be sensible for a party to alert the other to events that *might* occur as well as ones that *will* occur (if there is a reasonable probability). The aim of the warning is to allow a strategy to resolve or minimise the effect of the event to be agreed in advance of the event occurring, and some events might be entirely averted if a warning is issued sufficiently early. Whether or not a warning is given, the contractor is required to take reasonable steps to minimise the effect of any delaying event (cl 3.3.1).

Revisions to the contract completion date

4.24   All construction contracts include provisions for making revisions to the contract period (usually referred to as 'extensions of time', but in the RIBA Building Contracts they are referred to as 'Revisions of Time'). In many cases, construction contracts will list the reasons that would justify a revision, which will comprise 'neutral' events – i.e. things that could occur through neither party's fault, such as bad weather – and events caused by actions of the client or the contract administrator. Such a list of events acts as a means of distributing risk between the parties; the more events are included, the more risk is borne by the client. The main reason for including events caused by the client and the contract administrator is to preserve the client's right to liquidated damages. If no such provisions were included and a delay occurred that was caused by the client, this would in effect be a breach of contract by the client and the contractor would no longer be bound to complete by the completion date (see *Peak Construction v McKinney Foundations*). The client would therefore lose the right to liquidated damages, even if some of the blame for the delay rests with the contractor. The phrase 'time at large' is often used to describe this situation. However, this is, strictly speaking, a misuse of the phrase as in most cases the contractor would remain under an obligation to complete within a reasonable time.

Peak Construction (Liverpool) Ltd v McKinney Foundations Ltd (1970) 1 BLR 111 (CA)

Peak Construction was the main contractor on a project to construct a multi-storey block of flats for Liverpool Corporation. The main contract was not on any of the standard forms, but was drawn up by the Corporation. McKinney Foundations Ltd was the subcontractor nominated to design and construct the piling. After the piling was complete and the subcontractor had left the site, serious defects were discovered in one of the piles and, following further investigation, minor defects were found in several other piles. Work was halted while the best strategy for remedial work was debated between the parties. The city surveyor did not accept the initial remedial proposals, and it was agreed that an independent engineer would prepare a proposal.

The Corporation refused to agree to accept his decision in advance, and delayed making the appointment. Altogether it was 58 weeks before work resumed (although the remedial work took only six weeks) and the main contractor brought a claim against the subcontractor for damages. The Official Referee at first instance found that the entire 58 weeks constituted delay caused by the nominated subcontractor and awarded £40,000 damages for breach of contract, based in part on liquidated damages which the Corporation had claimed from the contractors. McKinney appealed, and the Court of Appeal found that the 58-week delay could not possibly be entirely due to the subcontractor's breach, but was in part caused by the tardiness of the Corporation. This being the case, and as there were no provisions in the contract for extending time for delay on the part of the Corporation, it lost its right to claim liquidated damages, and this component of the damages awarded against the subcontractor was disallowed. Even if the contract had contained such a provision, the failure of the architect to exercise it would have prevented the Corporation from claiming liquidated damages. The only remedy would have been for the Corporation to prove what damages it had suffered as a result of the breach.

4.25   In both CBC and DBC, the provisions are set out in clause 9.9, which states:

9.9   The Contractor may apply (with supporting documentation) for a Revision of Time if the Works are delayed due to any of the following:

9.9.1   the Architect/Contract Administrator issues a Change to Works Instruction

9.9.2   the Employer [CBC; Customer, DBC] defers access to the Site

9.9.3   the Employer [Customer] or its agents cause delay or disruption

9.9.4   the Architect/Contract Administrator postpones the Works or part of the Works

9.9.5   the Architect/Contract Administrator issues an instruction for any work or products to be inspected and/or opened up and no defective work or products are found

9.9.6   the Architect/Contract Administrator issues an instruction resolving an inconsistency, unless the inconsistency is due to a document prepared by the Contractor

9.9.7   the Architect/Contract Administrator issues instructions on Items of Interest

9.9.8   the Contractor suspends some or all of its duties

9.9.9   the action or omission of a utility company or statutory body

9.9.10   weather conditions are exceptionally adverse

9.9.11   the Employer's [Customer's] risks and/or Force Majeure.

4.26   Many of these events are dealt with at other points in this Guide: for change to works instructions, see paragraphs 5.21–5.30; for deferring access and postponing work, see paragraphs 4.6 and 4.17 above; for inspections and items of interest, see paragraphs

5.18–5.20; and for inconsistencies, see paragraphs 2.42–2.43. In regard to other events, the following points should be noted:

- The client or its agents causing delay or disruption (cl 9.9.3) acts as a useful 'catch-all', to pick up any actions not covered elsewhere in the list. 'Agents' would include the contract administrator if acting on behalf of the client, and may include other third parties whose actions have been authorised by the client.

- The contractor suspending some or all of its duties (cl 9.9.8) must be taken to be referring to a valid suspension, i.e. one allowed for under clause 8 (and not simply the contractor suspending an aspect of the work for other reasons, such as a disagreement with the client).

- Weather has to be exceptionally adverse (cl 9.9.10), i.e. not that which would be expected at the time of year in question. Any ambiguity here can usually be resolved by consulting the contractor's site records and/or Meteorological Office records.

- Client's risks are defined in clause 9.1 and force majeure in clauses 9.5 and 9.6 (both contracts), and are essentially matters for which the client has agreed to accept the risks of any delay (cl 9.2 and 9.8; the contractor accepts the risk of all other matters, cl 9.3 and 9.4). The list of risks does not specifically exclude matters that are caused by the contractor's negligence, for example delays caused by a fire, flood, etc., where the contractor has caused the fire etc. itself. This is similar to the position under the SBC11, where the employer accepts such risks. However, the RIBA has confirmed that in these contracts the clause 9.1 risks were not intended to include anything resulting from the contractor's negligence. As this intention is not apparent in the clause, it may be sensible for the client to consider clarifying it by an amendment (which should be set out in the tender documents).

Applying for a revision of time

4.27   If the contractor is delayed by a clause 9.9 event and wishes to apply for a revision of time, it must do so, with supporting information, within ten days of the 'single' event ending (cl 9.10.1). If the event is a continuing one, clause 9.10.2 states that 'the Contractor shall inform the Architect/Contract Administrator of the event within 10 days of it commencing and shall apply for a Revision of Time (with supporting documentation) within 10 days of the last element of the event'. An example of a 'single' event would be a pipe bursting overnight and causing damage; a 'continuing' event would be something that is ongoing, such as exceptionally adverse weather. The aim of clause 9.10.2 is to ensure that the contract administrator is made aware of the event at a reasonably early stage, even through the contractor might at that point not have enough information to make a full application. In the case of change to works instructions, an assessment is required within 10 days of *receiving* the instruction, not from when it first impacts on the programme, or when a possible impact is identified (cl 5.12 CBC and 5.9 DBC, see para 5.31), or the right to a revision will be lost.

4.28   The time limits are critical – and an important feature of the RIBA Building Contracts – as they limit the contractor's rights: clause 9.10.3 states: 'If the Contractor fails to apply within these periods, it will lose the right to a Revision of Time'. It not clear whether the contractor's application is a condition precedent to the award of a revision of time, i.e. not only would it lose the right, but the contract administrator would have no power to issue a revision unless the contractor has made an application. It is unlikely that the contractor would object to a revision made by the contract administrator on its own initiative, and given the 'discretion' afforded by clause 9.11.2 (see para. 4.31), it is unlikely that any such revision would cause problems. Nevertheless, it would be wise for the contract administrator to seek the agreement of both parties before making the revision.

4.29   The contract administrator and contractor are required to 'aim to agree the Revision of Time promptly' (cl 9.11.1, also 5.13.1 CBC and 5.10.1 DBC). If no agreement can be reached, the contract administrator is required to make a decision on the correct revision of time (cl 9.11.2, also 5.13.2 CBC and 5.10.2 DBC). If a revision of time is granted, the contract administrator is required to amend the date for completion, and to inform both parties (cl 9.12). This process seems reasonably straightforward; however, it is complicated by the fact that clause 9.11.2 in full states:

> If the Architect/Contract Administrator and the Contractor are unable to agree, *or the Contractor did not apply in time,* the Architect/Contract Administrator shall make a reasonable decision about the appropriate Revision of Time taking into consideration the Programme (if applicable) and any advance warning notices. [emphasis added]

4.30   At first sight there appears to be some contradiction between clauses 9.10.3 and 9.11.2: if the contractor has not met the time limits set, there should be no need for the contract administrator to go through the process of reaching a decision, the right to a revision has been simply lost.

4.31   One way of reading clause 9.11.2 is therefore that if the application is late, the only 'reasonable' decision would be that the revision of time is zero (and the purpose of issuing a 'decision' is simply to record the position). The RIBA has, however, suggested that clause 9.11.2 is intended to give the contract administrator discretion to consider the merits of the claim, even when the application is submitted late.

4.32   Clauses such as the above are often referred to as time-bar clauses, and tend to be interpreted strictly by the courts, especially when the matter that is the subject of the claim is a default of the client, or of those for whom the client is responsible, such as its consultants; the principle being that it is inherently unlikely that a party would relinquish its right to claim for losses suffered by the other party's breach. However, in this case the statement in clause 9.10.3 is clear and unequivocal.

4.33   It is suggested that if a claim is late, the contract administrator should normally apply the time bar, and resist the inevitable arguments that to do so is being 'unreasonable'. If there is any discretion, it is suggested that it is exercised with caution, and only in exceptional circumstances (e.g. where the delay has been raised earlier under an advance warning notice and discussed with the contractor, who has provided full details of its expected effects, but the actual application for a revision arrived slightly late). To treat it otherwise, i.e. to consider all applications, would deprive clause 9.10.3 of any meaning.

Assessment of an application

4.34   Assessment of revisions of time is a complex process that often causes difficulty in practice. What follows is a brief outline only. If faced with difficult claims the contract administrator should consult one of the published texts on the subject (e.g. Birkby et al.,

2008; Eggleston, 2009) or take expert advice.

4.35   All assessments must be made in a fair and reasonable manner. If the parties have agreed rules under optional clause A11 (CBC), A9 (DBC), then obviously these should be used for the assessment where appropriate. The objective is always to assess what effect the event will have on the final completion date, and the contract administrator should take into account the fact that the contractor should use reasonable endeavours to minimise the effects of any delaying event. This would include giving an early warning of any event of which it ought to be aware, and taking reasonable steps to re-organise the works and adjust the programme (cl 3.3.2).

4.36   It should be remembered that the standard of proof is 'on the balance of probability' (the civil standard), i.e. the contractor has to convince the contract administrator that it is more likely than not that it suffered delay due to the event. The contract administrator should not expect the application to prove the contractor's case 'beyond all reasonable doubt' (the criminal standard). The contractor must demonstrate not only that an event listed in clause 9.9 has occurred, but also that the event has delayed the work, and that the particular work delayed is on the critical path, i.e. its delay will ultimately delay the completion of the project; put simply, it must show a causal link between the event and delay to completion.

4.37   If there is a programme, the contract administrator is required to take it into account when making adjustments to the time (cl 5.13.2 in CBC, or cl 5.10.2 in DBC, and cl 9.11.2 in both versions). The contract administrator is also required to take into account 'any advance warning notices' (cl 9.11.2). There are probably two ways in which the notices may be relevant: first, the fact that they were given is evidence that the contractor has been vigilant and used reasonable endeavours to avoid the delay (conversely if no warning is given this must be taken into account, cl 3.3.2); and second, the notices should contain useful contemporaneous evidence of the nature and anticipated effects of the event. The contract administrator could also consult any other available records as to the history of events on site, and use its own knowledge and experience when making the decision.

4.38   Two issues in particular can cause problems when assessing revisions of time: 'concurrent delay' and 'contractor's float'. The following sections present very brief explanations of these issues.

Concurrent delay

4.39   Where two separate events contribute to the same period of delay, but only one of these is an event listed in clause 9.9, the normal approach is that the contractor is given a revision of time for the full effect of the clause 9.9 event (i.e. the contractor gets the benefit of the contributing but approximately equal cause, unless another competing cause can be identified as the dominant cause). The courts have normally adopted this approach (see *Walter Lilly & Co Ltd* v *Giles Mackay & DMW Ltd*; note, however, that the same does not apply to claims for loss/expense).

4.40   The instinctive reaction of many assessors might be to 'split the difference', given that both parties have contributed to the delay. However, it is more logical that the contractor should be given a revision of time for the full length of delay caused by the relevant event, irrespective of the fact that, during the overlap, the contractor was also causing delay.

Walter Lilly & Co Ltd v *Giles Mackay & DMW Ltd* [2012] EWHC 649 (TCC)

This case concerned a contract to build Mr and Mrs Mackay's, and two other families', luxury new homes in South Kensington, London. The contract was entered into in 2004 on the JCT Standard Form of Building Contract 1998 Edition with a Contractor's Designed Portion Supplement. The total contract sum was £15.3 million, the date for completion was 23 January 2006, and liquidated damages were set at £6,400 per day. Practical completion was certified on 7 July 2008. The contractor (Walter Lilly) issued 234 notices of delay and requests for extensions of time, of which fewer than a quarter were answered. The contractor brought a claim for, among other things, an additional extension of time. The court awarded a full extension up to the date of practical completion. It took the opportunity to review approaches to dealing with concurrent delay, including that in the case of *Henry Boot Construction (UK) Ltd* v *Malmaison Hotel (Manchester) Ltd* (where the contractor is entitled to a full extension of time for delay caused by two or more events, provided one is an event which entitles it to an extension under the contract), and the alternative approach in the Scottish case of *City Inn Ltd* v *Shepherd Construction Ltd* (where the delay is apportioned between the events). The court decided that the former was the correct approach in this case. As part of its reasoning the court noted that there was nothing in the relevant clauses to suggest that the extension of time should be reduced if the contractor was partly to blame for the delay.

Taking any other approach – for example, splitting the overlap period and awarding only half of the extension to the contractor – could result in the contractor being subject to liquidated damages for a delay partly caused by the client.

Float

4.41   Another area that sometimes causes problems is the question of float. Float is essentially planned early completion, i.e. a period shown on a programme between the contractor's planned completion date and the contractual date for completion. If a revision of time is applied for at a relatively early stage in the project, it may be that the delay suffered will not push the planned date beyond the contractual date for completion. Therefore, strictly speaking, no revision should be given. However, if the contractor is later delayed through its own errors, it may wish it had had the benefit of the earlier revision, as it now appears unlikely that it will complete on time. In such cases it is generally considered that the contractor should be given the benefit of the 'float', therefore the contract administrator may need to review earlier decisions and account for the float period.

Final assessment of revisions of time

4.42   The RIBA Building Contracts do not require the contract administrator to review any revisions of time after practical completion. However, the RIBA has confirmed that it would be possible to do this, provided that it extended, not reduced, the contract period (the RIBA proposes that the power arises under clause 5.4.8, but the author suggests that 9.11.2 is a better basis, see para. 4.28). After practical completion, the contract administrator will be able to review all the earlier revisions of time and to adjust the date for completion as necessary with the benefit of full information, including the final programme. Such an adjustment would only be to extend the date further; the contract administrator would not be entitled to shorten the programme at that stage.

5 Control of the works

5.1   The previous chapter focused on the programming of the works, and on monitoring progress in relation to the planned programme. This chapter examines the quality of the works: how it is achieved, who is responsible for it and what steps can and may need to be taken if problems are experienced.

5.2   Achieving the contractual standard is entirely the responsibility of the contractor, but the contract administrator also has a role to play, by providing information on the required standard and monitoring whether it is achieved. The RIBA Building Contracts also confer various powers on the contract administrator that can be used if the contract administrator feels it is necessary to step in.

Control of day-to-day activities

5.3   The day-to-day control of the works, i.e. the management of operations on site, co-ordination of orders and supplies, procurement of labour and subcontractors, and all issues relating to quality control, is entirely the responsibility of the contractor.

'Person-in-charge'

5.4   In order that its responsibility is carried out properly, the contractor is required to ensure that 'a suitably qualified representative is on Site during the Works to answer queries and receive instructions on its behalf' (cl 1.3). What constitutes 'suitably qualified' would depend on the nature and scale of the project, i.e. the representative should be sufficiently qualified to fulfil their role competently and in accordance with the contract. The contracts do not require that the person is present 'at all times' (as in the JCT's SBC11), and it may be that something less than full-time presence would be acceptable, provided that they are present at all material times, and that they make arrangements for dealing with queries or receiving instructions during short absences. There is no requirement in the Contract Conditions to have the person named, but it would be good practice to establish the identity of the person at the pre-start meeting, and to make sure this is recorded in writing.

Responsibility for subcontractors

5.5   The contractor is also fully responsible for the quality of work of all subcontractors (cl 1.7), whether these are its own domestic subcontractors or those selected by the client under the required specialists optional clauses (cl A7.1.3 CBC; or A5.1.3 DBC). As there are no provisions in the RIBA Building Contracts for the client to directly engage other firms, any persons on the site during the works would be under the direct supervision and responsibly of the contractor. If the client made any special arrangements with the contractor, outside of the contract, to allow its directly engaged persons on site, then this would cause confusion as to who is responsible for their performance, including in relation to quality of work, progress and health and safety, unless a detailed agreement regarding these matters is drawn up.

Principal designer

5.6   The RIBA Building Contracts require that both parties comply with all health and safety regulations (cl 4.3). The key relevant regulations are the Construction (Design and Management) Regulations 2015, which came into force on 6 April 2015 and apply to all construction projects.

5.7   For almost all projects, the Regulations require the client to appoint (in writing) a principal designer and a principal contractor (Regulation 5). The principal designer manages and co-ordinates health and safety aspects during the pre-construction phase, and then liaises with the principal contractor and co-ordinates ongoing design work during the construction phase. The principal designer could be the architect and/or the contractor administrator, but this is not necessarily the case; the role of principal designer is a distinct one and should normally be covered by a separate appointment. The principal contractor, which will almost always be the contractor under the contract, manages the construction phase of a project. This involves liaising with the client and principal designer throughout the project, including during the pre-construction phase, and producing a plan of how it will manage health and safety on site during the construction phase. If a domestic client fails to make the required appointments, the Regulations state that the designer in control of the pre-construction phase of the project is the principal designer, and that the contractor in control of the construction phase is the principal contractor (Regulation 7(2)).

5.8   If either party breaches its obligations under clause 4.3, it will be contractually liable to the other party, as well as liable under the Regulations. For a detailed understanding of their roles, the parties should consult the Regulations and related guidance.[1]

Flow of information

5.9   The most important function of the contract administrator is to ensure that the contractor is supplied with detailed and accurate information, either at tender stage or during the project, which makes clear precisely what standards and quality are required. In projects where the contractor is undertaking design, it will be required to submit its design to the contract administrator before or during the construction phase. The overall responsibility for integrating that design with the rest of the project rests with the contract administrator.

Information to be provided by the contract administrator

5.10   In most projects the information in the contract documents is not sufficient to construct the works; a certain amount of detailed information (e.g. schedules of finishes) is often outstanding, and it is usual for the contractor to be provided with this information during

---

[1]   See, for example, the Health and Safety Executive guidance note L153 *Managing Health and Safety in Construction*, and the CIC Risk Management Briefing *Construction (Design and Management) Regulations 2015*.

the course of the work. The RIBA Building Contracts – unlike, for example, the JCT's MW11 (cl 2·4) – do not contain an express obligation to provide further information. However, they do place a general duty on the contract administrator to issue instructions (cl 5.1), which would include instructions regarding detailed aspects of the works. In any case, such an obligation is likely to be implied with respect to all work that is not listed as a part to be designed by the contractor. It is difficult to see how a contractor could successfully achieve completion in the absence of a duty requiring that instructions, information, plans, drawings etc. are issued in good time, so the obligation is likely to arise under an implied duty to co-operate. The contract administrator should therefore assume that it should provide the contractor will all key information, the only exceptions being in relation to contractor-designed items and, possibly, very small items, where it may be that the contractor could be expected to determine these for itself (see *Wells* v *Army & Navy Co-operative Society Ltd*, also para. 3.13, and *National Museums and Galleries on Merseyside (Trustees of)* v *AEW Architects and Designers Ltd*).

> *Wells* v *Army & Navy Co-operative Society Ltd* (1902) 86 LT 764
>
> The Court of Appeal refused to allow the deduction of liquidated damages where late completion was partly caused by late provision of information by the architect (as well as by variations and a delay by the employer in giving possession). These were considered acts of prevention that were not catered for in the extension of time clause in the contract, and it was held that the liquidated damages provisions were ineffective and could not be applied to the delay period.

5.11   The time by which information should be supplied is whenever the contractor needs it, given the overall progress on site and the date for completion. The contractor's programme might have been required to set out dates for information to be provided, in which case this would be a guide, but it would not be conclusive. As with other issues concerning timing, it would be sensible to have this as an ongoing agenda item at progress meetings.

Information to be provided by the contractor

5.12   If optional clause A2 is selected, the contractor is required to provide information regarding its developing design, as set out in the Contract Details (item U in CBC, item R in DBC). The clause states:

> At least 21 days before carrying out any part of the Works listed under item U [CBC; item R DBC], the Contractor shall submit details of its design to the Architect/Contract Administrator for comment.

This is a very useful provision to include, as it is essential that the contract administrator is able to monitor that the design meets the client's requirements, as set out in the contract documents, and to co-ordinate it with the rest of the project.

5.13   The RIBA Building Contracts do not include a detailed procedure for submissions (e.g. format, response, comments or resubmission), such as that set out in Schedule 1 to SBC11. The administrator could, however, use the power under clause 5.4.7 to instruct that further or revised documents are provided. More importantly, the contracts do not include an equivalent clause to SBC11 Schedule 1:8·3, which states that:

neither compliance with the design submission procedure in this Schedule nor with the Architect/Contract Administrator's comments shall diminish the Contractor's obligations to ensure that the Contractor's Design Documents and CDP Works are in accordance with the Contract.

5.14   If an RIBA Building Contract is to be used for a project where there are significant contractor design elements, then it may be wise to consider including some provisions regarding submission. Failing that, when making comments it might be sensible to remind the contractor of its obligation to ensure that the design is in accordance with the client's specification (cl A2.1.3).

Inspection and tests

5.15   In addition to providing information, the contract administrator will also inspect the works at regular intervals to monitor whether the required standard is being met. The contract administrator may also, if necessary, issue instructions to have work opened up and tested, although this may have implications for the contract price and programme.

Inspection

5.16   On most projects the contract administrator will inspect the works at regular intervals. The RIBA Building Contracts do not place a duty on the contract administrator to do this, although they do give the contract administrator the power to visit the site and inspect the works (cl 5.3.1 and 5.3.2). Note that the reference in the guidance notes to the contract administrator's 'duty' to visit and inspect is incorrect. However the contract administrator's obligations to the client will, almost always, include a duty to inspect. Normally this would be an express duty under the terms of appointment, but in some circumstances it could also be implied: clearly, when the contract administrator is required under the contract to form an opinion on various matters – including being satisfied with the standard of work and materials prior to issuing a payment certificate – then it is essential that some form of inspection takes place. However, it is important to note that the duty is owed to the client, and not to the contractor. For example, a contractor cannot blame a contract administrator for failing to draw its attention to defective work.

5.17   Furthermore, a contract administrator will not necessarily be liable to the client for negligent inspection if a defect in a contractor's work is not identified. The question in every case is whether the contract administrator exhibited the degree of skill that an ordinary competent professional would exhibit in the same circumstances. Generally, the extent and frequency of inspections must enable the contract administrator to be in a position to properly certify that the construction work has been carried out in accordance with the contract (*Jameson v Simon*). The case of *McGlinn v Waltham Contractors Ltd* sets out some useful advice on the appropriate standard of inspection.

McGlinn v Waltham Contractors Ltd [2007] 111 Con LR 1

This case concerned a house in Jersey called '*Maison d'Or*' that was designed and built for the claimant, Mr McGlinn. The house took three years to build, but after it was substantially complete, it sat empty for the next three years while defects were investigated. It was completely demolished in 2005 having never been lived in, and was not rebuilt. Mr McGlinn brought an action against the various consultants, including the architect, and the contractor, claiming that *Maison d'Or* was so badly designed, and so badly built, that he was entitled to demolish it and start again. The contractor however had gone into administration and played no part in the hearing. The architect was engaged on RIBA Standard Form of Appointment 1982, which referred to 'periodic inspections'. HH Judge Peter Coulson QC usefully summarised the principles relating to inspection (at paras 215 and 218), which included the following:

- The change from 'supervision' to 'inspection' represented 'a potentially important reduction in the scope of an architect's services'.

- 'The frequency and duration of inspections should be tailored to the nature of the works going on at site from time to time'.

- 'If the element of the work is important because it is going to be repeated throughout one significant part of the building, then the inspecting professional should ensure that he has seen that element of the work in the early course of construction/assembly so as to form a view as to the contractor's ability to carry out that particular task'.

Testing and defective work

5.18   As noted at the beginning of this chapter, it is entirely the contractor's responsibility to ensure that the work is completed in accordance with the contract. The contract administrator is, however, given various discretionary powers that may be useful if it is concerned that the contractor does not appear to be fulfilling this primary obligation.

5.19   First, the contract administrator may issue instructions requiring any work to be uncovered, inspected and/or tested for compliance (cl 5.3.3, 5.4.3, and cl 5.7 in CBC or clause 5.5 in DBC). If the work proves to be defective, the contractor will bear the cost of the complying with the instruction and the correction of the defects (cl 5.7.1 or cl 5.5.1). If the work complies with the contract, the client will bear the cost of the complying with the instruction (cl 5.7.2 or cl 5.5.2). Generally, therefore, the contract administrator would only issue such an instruction if there was a serious concern, or if the failure of the element in question would be crucial to the project or extremely difficult to correct later. Failure to issue any instructions would not in any circumstances lessen the contractor's responsibility, no matter how difficult or expensive it might be to correct the problem later.

5.20   Second, whether or not the defective work has been tested, the contract administrator has power to instruct that it is removed (cl 5.4.3). Alternatively, the contract administrator may accept work that does not accord with the contract 'and adjust the Contract Price accordingly' (cl 5.8 in CBC, or cl 5.5.3 in DBC). Care should be taken when doing this. The contract administrator should obtain the client's agreement, and a value should be proposed and agreed (see para. 6.12). Although an instruction is not required (unlike in JCT contracts), it would be advisable for the contract administrator to confirm the acceptance and deduction in writing.

Contractor administrator's instructions

5.21   The RIBA Building Contracts give the contract administrator wide-ranging powers to issue instructions. In some cases these are expressed as being a duty, and generally if the contract administrator fails to issue instructions necessary for the progress of the works, this may constitute a breach by the client. The full list is set out in Table 5.1.

Table 5.1 Matters about which the contract administrator may issue an instruction	
Clause	**Instruction**
3.4.2	Changes required to effect cost savings and value improvements
5.1	General duty to issue instructions
5.4.1 and 5.12 (CBC) or 5.9 (DBC)	Change to works instructions
5.4.2	Instructions for dealing with items of interest found on the site
5.4.3	Instructions on tests or inspections
5.4.3	Instructions rejecting defective work or products
5.4.4	Instructions postponing the works or sections of the works
5.4.5	Instructions to remove persons from the works
5.4.6	Instructions to resolve inconsistencies in the contract documents or in a previous instruction
5.4.7	Instructions requiring further documents
5.4.8	Instructions on any clause to enable good administration
5.6.4 (CBC) or 5.7.4 (DBC)	Instruction that oral instructions shall not take effect until confirmed in writing by the contract administrator
5.7 (CBC) or 5.5 (DBC)	Instruction requiring work is uncovered and inspected or tested
5.14 (CBC) or 5.11 (DBC)	Instruction dealing with inconsistencies in contract documents
9.7	Instruction to stop work if force majeur occurs
11.11.2	Instructions on communication procedures

5.22   There is some overlap between clause 5.4.8 and the other clauses in Table 5.1. However, it serves as a 'catch-all' and gives the contract administrator a wide discretionary power.

5.23   Note that a 'Change to Works Instruction' is one that alters the 'design, quality and/or quantity' of the works. This is a narrower definition than in JCT contracts, in that it appears not to cover the manner of carrying out the works, working hours, access to the site and general management matters. However, the clause 5.4.8 provision for 'Instructions on any clause of the Contract to enable good administration' may well cover most of these matters.

Delivery of instructions

5.24   All instructions are required to be in writing, and the contractor is required to comply with them immediately (cl 5.5 in CBC, or cl 5.6 in DBC). If the contract administrator gives an instruction orally, the contract requires that it is confirmed in writing 'promptly' (cl 5.6.1/ cl 5.7.1). If the contract administrator does not confirm the instruction, the contractor 'shall issue a written record of the oral instruction to the Architect/Contract Administrator' (cl 5.6.2/cl 5.7.2). Clause 5.6.3/5.7.3 then states 'Except where clause 5.6.4/5.7.4 applies, a written record under clause 5.6.2/5.7.2 shall be the record of the Architect/Contract Administrator's instruction unless the Architect/Contract Administrator amends it in writing within 5 days of receiving it'.

5.25   Ideally, contract administrators should avoid giving oral instructions, except in cases of emergency. If they cannot be avoided, it should not be difficult to confirm an instruction promptly using electronic communications. Remember that a simple email would constitute an instruction in writing; there is no need (although it may be good practice) to issue it in any special format.

5.26   The contracts (cl 5.6.2 in CBC and cl 5.7.2 in DBC) refer to the common practice whereby the contractor issues a 'written record of the oral instruction' (often termed a 'confirmation of verbal instruction' or 'CVI'). Many contract administrators are comfortable with having the contractor confirm what they have said, but if this system is used, the contract administrator must be very vigilant and check that a contractor's confirmation exactly reflects what was intended. In a busy office with a constant stream of emails, it is easy to misread one, and an inaccurate confirmation will become the contractual record after five days.

5.27   If the contract administrator would prefer to avoid this situation, it can instruct that oral instructions do not take effect until confirmed by the contract administrator in writing (cl 5.6.4 in CBC, cl 5.7.4 in DBC). The contractor is still required to send the written record under clause 5.6.2/6.7.2.

5.28   In the first system for dealing with oral instructions, it appears that the instruction is intended to take effect immediately; although the contracts do not say this, clauses 5.6.2 and 5.6.3 of CBC and clauses 5.7.2 and 5.7.3 of DBC refer to a record of an instruction, as if it is already effective – it simply needs to be recorded. However, under the second system, an oral instruction would be of no effect until confirmed by the contract administrator. Although this system would prevent inaccurate CVIs slipping through the net, there is a danger that an instruction could get overlooked entirely. The choice of which system to use is a matter of personal preference for the contract administrator. It can be decided after the contract is entered into, but the decision should be made early on in the project, ideally at the pre-contract meeting, as otherwise there could be confusion if an emergency unexpectedly arises.

Procedure following an instruction

5.29   In CBC, the contractor is given the power to notify the contract administrator, within seven days of receiving an instruction, if it believes the instruction is not in accordance with the contract or that implementing it would have adverse health and safety implications or would adversely affect any part of the works designed by the contractor (cl 5.9). (It is suggested that, in order to discharge its duty of care, the contractor should notify the contract administrator if any of these situations arise; this duty would apply to both contracts.) The contract administrator may then, on receipt of the notification, modify, amend, withdraw or confirm the instruction, and the contractor shall comply accordingly (cl 5.10).

5.30   In both contracts, if the contractor fails to comply with an instruction, the contract administrator may issue the contractor with a seven-day notice to comply (cl 5.11 in CBC, cl 5.8 in DBC), and if the contractor fails to comply with the notice, the client may engage others to undertake the instruction (cl 5.11.1/cl 5.8.1); this works in a very similar way to the notice to comply provisions in JCT contracts. As with the JCT notice, if the contract administrator has serious concerns that the contractor may refuse to comply (e.g. because it has already expressed that intention at a meeting) then it would be possible for it to issue the instruction and the compliance notice together. In addition to giving the client this right, the contracts require the contractor to co-operate with the new contractors (cl 5.11.2/cl 5.8.2) and to allow them access to the site (cl 5.11.3/cl 5.8.3), and the contractor is to be responsible for all costs and expenses incurred by the client (cl 5.11.4/cl 5.8.4).

5.31   Where a change to works instruction is issued, the contractor is required to calculate and submit details to the contract administrator of its effect on the contract price and date for completion (cl 5.12 CBC, 5.9 DBC). This must be done within 10 days of receiving the instruction. The contract administrator and contractor should aim to agree the appropriate revision of time and/or additional payment promptly, otherwise the contract administrator determines the appropriate amount (cl 5.13 CBC, 5.10 DBC).

Practical completion

5.32   The decision to certify practical completion is one of the most important that the contract administrator makes during the whole project as it triggers many contractual consequences that are important to both the client and the contractor. The period leading up to practical completion can be difficult and stressful. Sometimes this may be due to an (erroneous) belief by the contractor that the word 'practical' indicates that something that is 90 per cent finished, or could be occupied, has reached practical completion. At other times there may be pressure from a client who is very anxious to occupy the building, and who may not appreciate how much work is still needed to correct what might appear to minor matters. However, in the RIBA Building Contracts (unlike many other contracts), a clear definition of practical completion is given. Clause 9.15 states:

> For Practical Completion to occur, the following must apply:
>
> 9.15.1 no aspect of the Works or a Section of the Works shall be outstanding
>
> 9.15.2 the Works shall be uncluttered and safe
>
> 9.15.3 any requirements stated in the Contract Documents or required by law for Practical Completion shall have been satisfied.

5.33   In addition, if optional clause A8 of CBC is selected, the contractor must have provided copies of all collateral warranties/third party rights agreements required before practical completion can be certified (cl A8.2).

5.34   It is suggested that 'no aspect of the Works or a Section of the Works shall be outstanding' includes aspects of quality, as well as quantity, therefore if the quality of work is unsatisfactory, practical completion as defined has not been reached. This interpretation has been supported by the courts, for example in the well-known case of *H W Nevill (Sunblest) Ltd* v *William Press & Son Ltd*. If the client nevertheless wishes to occupy the building with minor work outstanding, then a special arrangement will need to be made, as discussed below (see para. 5.40).

H W Nevill (Sunblest) Ltd v *William Press & Son Ltd* (1981) 20 BLR 78

Here William Press entered into a contract with Sunblest to carry out foundations, groundworks and drainage for a new bakery on a JCT63 contract. A practical completion certificate was issued, and new contractors commenced a separate contract to construct the bakery. A certificate of making good defects and a final certificate were then issued for the first contract, following which it was discovered that the drains and the hard standing were defective. William Press returned to the site and remedied the defects, but the second contract was delayed by four weeks and Sunblest suffered losses as a result. It commenced proceedings, claiming that William Press was in breach of contract and in its defence William Press argued that the plaintiff was precluded from bringing the claim by the conclusive effect of the final certificate. Judge Newey decided that the final certificate did not act as a bar to claims for consequential loss. In reaching this decision he considered the meaning and effect of the certificate of practical completion and stated (at page 87): 'I think that the word "practically" in clause 15(1) gave the architect a discretion to certify that William Press had fulfilled its obligation under clause 21(1) where very minor *de-minimis* work had not been carried out, but that if there were any patent defects in what William Press had done then the architect could not have issued a certificate of practical completion.'

5.35   It is open to the parties to set out their own particular requirements for practical completion, including the standard expected and the means for establishing if it has been reached (cl 9.15.3). The parties might consider, for example, requiring that mechanical services are properly commissioned, that any performance in use criteria are tested and checked (e.g. airtightness and acoustic requirements), that an operations manual is handed over and the client trained in operation of the building. Any of these would, of course, have to have been made clear in the tender documents.

5.36   The contractor is required to notify the contract administrator when it thinks that practical completion of the works, or a section, has been achieved (cl 9.16). If the contract administrator agrees, it will issue a certificate of practical completion of the works or section as appropriate (cl 9.16.1).

5.37   If the contract administrator does not agree, it must inform the contractor of this (cl 9.16.2). It is important to note that the contract administrator is not required to give reasons for this decision, and there is certainly no obligation to produce 'snagging lists'. It is common practice for the contract administrator to issue such a list, but, unless it has entered into special terms of appointment with the client, there is no need for it to do so. Not only is it very time consuming and resource hungry, it is effectively taking on the contractor's quality assurance duties, which could ultimately lead to a confusion as to roles and responsibilities. If the contract administrator is concerned about particular defects, there is no harm in informing the contractor, provided it is made clear that these are just examples of some of the shortfalls on the project and that they are not intended to be a comprehensive list.

Consequences of practical completion

5.38   The consequences of practical completion are as follows:

- liquidated damages will cease (cl 10.1);
- half the withheld retention is released (cl 7.4 CBC, cl 7.2 DBC);
- the defects fixing period commences (cl 10.2);
- the contractor remedies defects identified during the defects fixing period (cl 10.3.1; there is no requirement for the contract administrator to give notification of defects);
- the client allows the contractor reasonable access to remedy defects (cl 2.1.3);
- due dates change from monthly to two-monthly (cl 7.1, CBC only).

5.39   These consequences are important and therefore the contract administrator should take great care to ensure the defined level of completeness has been reached before certifying practical completion. Issuing the certificate with work outstanding places considerable additional risk on the client: the key areas being that there is no longer the sanction of liquidated damages to encourage the contractor to finish promptly, and the client holds only half the retention sum as security against any hidden defects. In addition, matters such as insurance of the works and health and safety will need to be resolved, and there are practical issues to do with managing the programming and payment for the outstanding work. The contracts have no provisions to cover these aspects as it assumes that the work will be finished, with the exception of defects that appear during the defects fixing period (see para. 5.46).

Use/occupation before practical completion

5.40   If the works have not reached practical completion, but the client wishes to use them, the contracts contain two provisions that may be of help to the client. The first is clause 9.17, which states that the client may request to use part of the works or a section of the works for 'storage or other purposes'. In theory at least, the clause places no limits on what form the use might take. The contractor must grant permission, but only if the use does not interfere with the carrying out of the works. If the contractor agrees, it would be wise for the contract administrator to confirm this in writing, so that later this use cannot be raised as a reason for claiming a revision of time or additional payment. Practical completion is not certified and therefore none of the above consequences apply, although the contracts state that the client becomes responsible for the insurance implications of such use (cl 9.17).

5.41   The second provision allows the client to request to 'take over' (as opposed to 'use') any part or parts of the works or a section of the works before the contract administrator certifies practical completion (cl 9.18). As above, the contractor must grant permission, but only if the use does not interfere with the carrying out of the works. If the contractor agrees, the contract administrator must issue a notice clearly identifying the areas to be taken over and the date of takeover (cl 9.18.1). The contracts then state (cl 9.18.2):

> The part(s) in the Works/Section(s) identified in the Architect/Contract Administrator's notice shall be viewed as having achieved Practical Completion. The Defects Fixing Period for the relevant part(s) shall start from the date of takeover.

5.42   Although it is not entirely clear, the reference to the part or section being 'viewed as having achieved Practical Completion' suggests that (a) the works to that part or section should be complete before takeover and (b) all the consequences of practical completion are intended to be triggered by the taking over, not just the defects fixing period. For example, although there is no specific reference to any reduction in liquidated damages, it would seem reasonable for some reduction to be made; this could be calculated on the basis of the proportion of the value of the completed works to the contract price. A similar reduction could be applied to the withheld retention. In order to avoid any doubt, the parties would be wise to agree the consequences in advance of the takeover, and the contract administrator to confirm them in the takeover notice.

5.43   There can be situations where the client is anxious to occupy part or possibly the whole of the works before any parts are sufficiently complete to be taken over under clause 9.18.

Rather than viewing the work as having reached practical completion when it has not, it would be better if the parties make an ad hoc agreement as to what the arrangement will be. A suggestion was put forward in the 'Practice' section of the *RIBA Journal* (February 1992) which has frequently proved useful in practice: in return for being allowed to occupy the premises, the client agrees not to claim liquidated damages during the period of occupation. Practical completion obviously cannot be certified, and the defects fixing period will not commence, nor will there be any release of retention money, until the work is complete. Health and safety will need to be given careful consideration, and matters of insuring the works will need to be settled with the insurers.

5.44   Because such an arrangement would be outside the terms of the contract, it should be covered by a properly drafted agreement that is signed by both parties. (The cases of *Skanska* v *Anglo-Amsterdam Corporation* and *Impresa Castelli SpA* v *Cola Holdings Ltd* illustrate the importance of drafting a clear agreement.) It may also be sensible to agree that, in the event that the contractor still fails to achieve practical completion by the end of an agreed period, liquidated damages would begin to run again, possibly at a reduced rate. In most circumstances this arrangement would be of benefit to both parties, and is certainly preferable to issuing a heavily qualified takeover notice, or a certificate of practical completion, listing numerous incomplete items of work.

> *Skanska Construction (Regions) Ltd* v *Anglo-Amsterdam Corporation Ltd* (2002) 84 Con LR 100
>
> Anglo-Amsterdam Corporation (AA) engaged Skanksa Construction (Skanska) to construct a purpose-built office facility under a JCT81 With Contractor's Design form of contract. Clause 16 had been amended to state that practical completion would not be certified unless the certifier was satisfied that any unfinished works were 'very minimal and of a minor nature and not fundamental to the beneficial occupation of the building'. Clause 17 of the form stated that practical completion would be deemed to have occurred on the date that the employer took possession of 'any part or parts of the Works'. AA wrote to Skanska confirming that the proposed tenant for the building would commence fitting-out works on the completion date. However, the air-conditioning system was not functioning and Skanska had failed to produce operating and maintenance manuals. Following this date the tenant took over responsibility for security and insurance, and Skanska was allowed access to complete outstanding work. AA alleged that Skanska was late in the completion of the works and applied liquidated damages at the rate of £20,000 per week for a period of approximately nine weeks. Skanska argued that the building had achieved practical completion on time or that, alternatively, partial possession of the works had taken place and that, consequently, its liability to pay liquidated damages had ceased under clause 17.
>
> The case went to arbitration and Skanska appealed. The court was unhappy with the decision and found that clause 17·1 could also operate when possession had been taken of all parts of the works and was not limited to possession of only part or some parts of the works. Accordingly, it found that partial possession of the entirety of the works had, in fact, been taken some two months earlier than the date of practical completion, when AA agreed to the tenant commencing fit-out works. Consequently, even though significant works remained outstanding, Skanska was entitled to repayment of the liquidated damages that had already been deducted by AA.

Impresa Castelli SpA v *Cola Holdings Ltd* (2002) CLJ 45

Impresa agreed to build a large four-star hotel for Cola Holdings Ltd (Cola), using the JCT Standard Form of Building Contract with Contractor's Design, 1981 edition. The contract provided that the works would be completed within 19 months from the date of possession. As the work progressed, it became clear that the completion date of February 1999 was not going to be met, and the parties agreed a new date for completion in May 1999 (with the bedrooms being made available to Cola in March) and a new liquidated damages provision of £10,000 per day, as opposed to the original rate of £5,000. Once the agreement was in place, further difficulties with progress were encountered, which meant that the May 1999 completion date was also unachievable. The parties entered into a second variation agreement, which recorded that access for Cola would be allowed to parts of the hotel to enable it to be fully operational by September 1999, despite certain works not being complete (including the air conditioning). In September 1999, parts of the hotel were handed over, but Cola claimed that such parts were not properly completed. A third variation agreement was put in place with a new date for practical completion and for the imposition of liquidated damages. Disputes arose and, among other matters, Cola claimed for an entitlement for liquidated damages. Impresa argued that it had achieved partial possession of the greater part of the works, therefore a reduced rate of liquidated damages per day was due. The court found that, although each variation agreement could have used the words 'partial possession', they had in fact instead used the word 'access'. The court had to consider whether partial possession had occurred under clause 17·1 of the contract, which provides for deemed practical completion when partial possession is taken, or whether Cola's presence was merely 'use or occupation' under clause 23·3·2 of the contract. The court could find nothing in the variation agreements to suggest that partial possession had occurred. It therefore ruled that what had occurred related to use and occupation, as referred to in clause 23·3·2 of the contract, and that the agreed liquidated damages provision was therefore enforceable.

Non-completion

5.45   If the contractor fails to achieve practical completion of the works by the date for completion, the client may deduct liquidated damages at the agreed rate (cl 10.1). There is no need for the contract administrator to have issued any certificate, although it would normally write to the client advising it of the position.

The defects fixing period

5.46   The contractor is required to remedy defects identified during the 'Defects Fixing Period' (cl 10.3.1). The period begins at practical completion and lasts for the length of time entered in the Contract Details (cl 10.2, Item K CBC and J DBC), the default period being 12 months (12 months is commonly used so that any mechanical services such as heating systems can be run through four seasons). A shorter period might be acceptable for very small projects.

5.47   The contracts do not state who should identify the defects. As part of its general duty to complete the works satisfactorily, the contractor would be obliged to ascertain whether any defects have appeared. There is no requirement for the contract administrator to prepare a schedule of defects at the end of the defects fixing period, as there is in some of the other standard contracts (e.g. IC11, cl 2·30), and the contractor's obligation to correct work does not depend on it having been notified by the contract administrator. However, the contract administrator is required to issue a notice to the contractor requiring it to remedy any defect *which it fails to fix* (cl 10.3.3). If the contractor does not comply with the notice promptly, the client may engage others to rectify the problem, and all costs are the responsibility of the contractor (cl 10.3.4). This right would only arise if the correct procedures are followed; if the client, for example, refuses to allow the contractor reasonable access for inspecting and undertaking work, this might result in the client being unable to claim the costs of remedying the defects by others (*Pearce and High Ltd v John P Baxter and Mrs A S Baxter*).

5.48   In practice, correction of defects is normally left until the end of the defects fixing period, as it is usually more convenient for both parties if all the work is done together. However, sometimes a defect can cause considerable problems to the client, in which case the contractor should take steps as soon as it is aware of the matter. In either case, if the contractor fails to deal with any defect satisfactorily, the contract administrator should issue a clause 10.3.3 notice, in order to trigger the client's right to engage others if necessary. After the end of the defects fixing period, when the contract administrator is satisfied that all defects are remedied, the contract administrator is required to notify the parties accordingly (cl 10.3.2).

6 Payment and certification

6.1 In any building contract, the key obligation of the client is to ensure the contractor is paid according to the contract, both in terms of the amounts paid and the timing of these payments. To pay the contractor less than it is rightfully due or to pay the right amounts but later than agreed will put financial strain on the contractor (normally through increased borrowing costs), which in extreme cases could result in bankruptcy. On the other hand, paying too much, or too early, will put the client at risk; should the contractor repudiate the contract it will be difficult, and in some cases impossible, to recover the money.

6.2 The RIBA Building Contracts contain detailed provisions concerning the appropriate amounts due to the contractor, how and when these amounts are to be assessed, when they become due and the procedures for payment, all of which are discussed below.

6.3 Both contracts offer a choice between monthly certification based on the value of work completed, a single interim payment following practical completion, or milestone payments. In addition, CBC offers an option of making an advance payment.

6.4 CBC also includes provision for the contractor to make applications for interim payments, and the timing and detail of the procedure to be followed are more complex. As this is an area where CBC and DBC differ from one another in several respects (see Table 6.1), separate sections are used to cover certification and payment for each version.

Table 6.1 Payment clause comparison

CBC		DBC	
Clause	Title/matter covered	Clause	Notes
Interim payments			
7.1	Due dates (does not apply if item W2 selected)	–	*No equivalent*
7.2	Contractor application before due date (does not apply if item W2 selected)	–	*No equivalent*
7.3	Issue of payment certificate, coverage	7.1	*Shorter, a simpler procedure*
7.4	Retention percentage	7.2	*Identical except for formatting*
		7.3	*Invoice, different to CBC cl 7.8*
7.5	Failure to issue a payment certificate	7.4	*Contractor's application following failure, a simpler procedure than CBC cl 7.2, 7.5 and 7.6*
7.6	Contractor's right to issue a payment notice	7.5	*Issue of payment certificate following application*

Table 6.1 Payment clause comparison – Continued

CBC Clause	Title/matter covered	DBC Clause	Notes
7.7	Employer's obligation to pay	–	*No equivalent*
7.8	VAT invoice	7.7	*Similar, reference to payment notice not included*
Pay less notice			
7.9	Pay less notice	7.6	*Significant difference: customer to inform contractor at time of making deduction*
7.10	Pay less notice	–	*No equivalent*
Final contract price			
7.11.1	Contractor's calculation	7.8.1	Same
7.11.2	Agreement of price	7.8.2	Same
7.11.3	Failure to submit or agree	7.8.3	Similar
		7.8.4	*Contractor's application*
7.12	Due date	–	*No equivalent*
Final payment			
7.13	Final payment certificate or notice	–	*No equivalent*
7.14	Final date for payment	7.9.1	*Simpler procedure*
7.15	Payment of payment certificate or notice	7.9.1	*Simpler procedure*
Other provisions			
8.1	Suspension	8.1	Same
8.2	Right to interest	–	*Not included*
8.3	Entitlement to costs	8.2	Same
9.13–9.14	Additional payment	9.13–9.14	Same
Optional clauses			
A4	Payment on completion of the works and milestone payments	A4	*Different, milestone payments only, simpler system*
A5	Advanced payment	–	*Not included*
A6	Evidence of ability to pay the contract price	–	*Not included*
A11	Rules for valuation of revision of time and additional payment	A9	Same

The contract price

6.5   The 'Contract Price' is defined in both contracts as 'the amount that the Employer/ Customer shall pay the Contractor for carrying out and completing the Works in accordance with the Contract'. The Agreement states 'the Employer/Customer shall pay the Contractor the Contract Price, which will be calculated in accordance with the Contract'.

6.6   The Contract Price is set out in item O of the Contract Details of both contracts. This can be either a fixed amount (a sum is inserted) or an amount calculated in accordance with a schedule of rates. In the latter case, the amount due will be calculated in relation to the quantity of work actually carried out, based on the schedules of rates and prices provided by the contractor at tender. However, even if 'fixed', the contract price will be subject to change, as a result of any change to works instructions, claims for additional payment and other mechanisms, as discussed below.

Adjustments to the contract price

6.7   References to adjustments to the contract price can be found in the following clauses:

- (cl 3.4) contractor's proposed changes that will improve quality and/or reduce the contract price;
- (cl 5.7, CBC; cl 5.5, DBC) costs resulting from instructions regarding tests, etc. being added to the contract price;
- (cl 5.8, CBC; 5.5.3, DBC) adjustment to the contract price following the contract administrator's acceptance of non-conforming work;
- (cl 5.12, CBC; 5.9, DBC) adjustments due to change to works instructions;
- (cl 9.13) 'an event attributable to the Employer [Customer] or its agents' adding costs and expenses to the works, entitling the contractor to apply for an adjustment (additional payment).

6.8   There are no 'fluctuations' clauses in either version of the contract, therefore the fact that the price of materials or labour may have changed since the contractor tendered would not be reason for the contract price to change.

Contractor's proposed changes

6.9   As part of the Collaborative Working section, the contractor is encouraged to propose changes to the works that will improve quality and/or reduce the contract price (cl 3.4). The clause then states that if such proposals are made, the client may seek the advice of the contract administrator and/or accept the proposed changes and have the contract administrator issue the necessary instructions to implement them. The wording of this clause may be of some concern to the contract administrator, as it certainly implies that the matter could be agreed between the contractor and the client before the contract administrator learns of it. It is suggested that the client would be wise to always consult the contract administrator before any proposals are accepted, in case they have adverse affects on other aspects of the design or result in increased costs in other aspects of the project.

6.10   In any event, the accepted proposals will ultimately be covered by an instruction, which ought to confirm the exact details of the change and the reduction to the contract price. Clause 3.5 states that any cost saving is to be divided equally between the parties. The contractor should, therefore, show in its proposal exactly how any savings are calculated. If there is any subsequent negotiation of the savings, the final reduction should be recorded before the instruction is issued.

Costs resulting from instructions regarding tests, etc.

6.11   Under clause 5.7 in CBC or clause 5.5 in DBC, the contract administrator may instruct that work is uncovered or tested (see paras. 5.18–5.20). If the work turns out to be in accordance with the contract, the costs resulting from the instruction are to be added to the contract price. The costs would include not only those of the uncovering and the tests themselves, but also for reinstating the work, making good any damage and disruption to general progress. If these are likely to be significant, the contract administrator could ask the contractor for an assessment of costs before issuing the instruction, in order to fully inform the client regarding the risks and benefits of the instruction prior to proceeding.

Acceptance of non-conforming work

6.12   As noted in paragraph 5.20, the contract administrator has the power to accept work that does not accord with the contract 'and adjust the Contract Price accordingly' (cl 5.8, CBC; cl 5.5.3, DBC). Care should be taken when doing this. The contract administrator should obtain both parties' agreement, and ensure that its proposed adjustment to the contract price is agreed and confirmed in writing. The client may not be happy to accept the work (or may later forget that it agreed to it). The contractor may prefer to correct the work, especially if the proposed reduction in the contract price is significant, and in general it cannot be denied the opportunity to do so (*Mul* v *Hutton Construction Limited*).

> *Mul* v *Hutton Construction Limited* [2014] EWHC 1797 (TCC)
>
> This case concerned what constitutes an 'appropriate deduction' when an employer decided to accept non-conforming work. Although decided in relation to a JCT contract (the JCT Intermediate Building Contract 2005), it is nevertheless applicable to the RIBA Building Contracts. The project concerned an extension and refurbishment work to a country house. A practical completion certificate was issued with a long list of defects attached, and during the defects liability period the employer decided to have this work corrected by other contractors. The employer then started court proceedings against the contractor, to claim back the costs of this work.
>
> A key issue was the interpretation of clause 2·30, which provides that the contract administrator can instruct the contractor not to rectify defects and 'if he does so otherwise instruct, an appropriate deduction shall be made from the Contract Sum in respect of the defects, shrinkages or other faults not made good'. In this case the contractor argued that an 'appropriate deduction' was limited to the relevant amount in the contract rates or priced schedule of works. The court disagreed. It decided that 'appropriate deduction' under clause 2.30 meant 'a deduction which is reasonable in all the circumstances', and could be calculated by any of the following: the contract rates or priced schedule of works; the cost to the contractor of remedying the defect (including the sums to be paid to third party subcontractors engaged by the contractor); the reasonable cost to the employer of engaging another contractor to remedy the defect; or the particular factual circumstances and/or expert evidence relating to each defect and/or the proposed remedial works.
>
> However, the judge also pointed out that the employer will still have to satisfy the usual principles that apply to a claim for damages, which include showing that it mitigated its loss. If the employer unreasonably refused to let the contractor rectify defects, then it is likely to find its damages limited to what it would have cost the contractor to put them right.

Change to works instructions

6.13   The process for assessing an adjustment to the contract price as a result of a change to works instruction is covered by clause 5.12 in CBC and clause 5.9 in DBC:

> Within 10 days of receiving a Change to Works Instruction, the Contractor shall calculate the effect (if any) of the instruction on the Contract Price and/or the Date for Completion, and submit details to the Architect/Contract Administrator.

The clause goes on to state that 'After this period the right to a Revision of Time and additional payment will be lost'. Just below this (cl 5.13.2, CBC and/or in DBC cl 5.10.2, DBC), the contract states that if the contractor fails to submit the calculation within the time period, or the contract administrator and contractor are unable to agree the revision/payment, 'the Architect/Contract Administrator shall determine the appropriate adjustment'. This requirement is not discretionary, but imperative: the contract administrator must make this determination.

6.14   There are therefore two apparently conflicting statements that apply when the contractor submits its calculation late: under clause 5.12 (CBC) and clause 5.9 (DBC) the contractor loses the right to any additional payment, but under clause 5.13.2 (CBC) and clause 5.10.2 (DBC), the contract administrator is required to assess how much it should be paid.

6.15   As noted in relation to the equivalent 'time-bar' clauses relating to revisions of time (see paras, 4.31–4.33), given the clear wording of clause 5.12 in CBC and clause 5.9 in DBC it is suggested that only in exceptional circumstances should the contract administrator decide that the value of the late application is anything other than zero.

Additional payment

6.16   Clause 9.13 (both contracts) is headed 'Additional Payment' and states:

> If an event attributable to the Employer/Customer or its agents adds costs and expenses to the Works, the Contractor may apply for an adjustment to the Contract Price.
>
> 9.13.1 If the event is a single event, the Contractor shall apply for the additional payment within 10 days of the event ending.
>
> 9.13.2 If the event is a continuing event, the Contractor shall inform the Architect/Contract Administrator of the event within 10 days of it commencing and shall apply for the additional payment (with supporting documentation) within 10 days of the last element of the event.
>
> 9.13.3 If the Contractor fails to adhere to these time periods, the right to the additional payment will be lost.

It should also be noted that in the case of additional costs due to change to works instructions, an assessment is required within 10 days of *receiving* the instruction (cl 5.12 CBC, cl 5.9 DBC see para 5.31), or the right to a revision will be lost.

6.17   The 'additional payment' clauses are intended to be a means of dealing with what is usually referred to as 'loss and/or expense'. A clause 9.13 'event' is not defined in the RIBA

Building Contracts (unlike JCT contracts, there are no listed 'relevant matters'); however, it is suggested that this term is wide enough to include any action by the client or its agents (which would include the contract administrator), including the issue of a change to works instruction. There is, therefore, some possible overlap between clause 9.13 and clause 5.12 (CBC) or clause 5.9 (DBC). However, in practice the contractor is likely to apply for all the consequences of a change to works instruction together (as the time limits are the same) and, as long as the contract administrator takes care not to duplicate any award for losses, there should be no difficulty. A similar procedure is followed regarding applications to that for revisions of time, and for change to works instructions, i.e. the contract administrator and contractor endeavour to agree, otherwise the administrator determines the appropriate amount (cl 9.14, see paras. 4.31–4.33 and 6.13–6.15 above). If the parties have agreed rules under optional clause A11, then obviously these should be used for the assessment where appropriate.

Certification and payment – CBC

6.18   As noted above, CBC includes several payment options. The default system is the usual one of payment at monthly intervals, but there is also an optional clause (cl A4) providing for a single payment on practical completion or for milestone payments. There is a further optional clause (A5) providing for advanced payment.

6.19   All procedures for payment in CBC are intended to comply with those set out in the Housing Grants, Construction and Regeneration Act 1996 (as amended) (the Housing Grants Act); for a full explanation of this Act see paragraphs 1.14–1.19.

Advanced payment

6.20   Optional clause A5 provides for the employer to make a payment in advance of the start date. Item X of the Contract Details requires the parties to insert the amount, the date for payment and a schedule of repayment instalments, giving dates and amounts. It also provides an option for requiring an advance payment security. This would normally be in the form of a bond, and clause A5.2 states that the employer is not liable for any advance payment until 'the Contractor has met the stated requirements for the Advanced Payment bond'. It is suggested that the advanced payment provisions are only used with a security, as making an advanced payment to the contractor is a significant risk for the employer (e.g. the contractor could become insolvent before the monies are expended on the employer's behalf). Although many contractors will argue that they are required to pay out large amounts in advance to their subcontractors and suppliers, and are therefore out of pocket, it is generally better that they should take this risk. If a significant reduction in price is offered to the employer, the employer should consider the risks carefully before making a decision.

Interim payments – monthly certification

6.21   The dates for both certification and payment are related to 'Due Dates' (the term 'Due Date' derives from the Housing Grants Act, which requires all contracts to establish due dates for payment). The first due date is stated in the Contract Details (item P) and subsequent due dates are the same day on each subsequent month (cl 7.1). If no first due date is inserted, the default is 30 days after the start date. It should be noted that altering these periods, i.e. including a longer or shorter period before the first due date or increasing or reducing the intervals, would not constitute a breach of the Act, provided there were still regular payment dates.

6.22   The contractor is entitled to submit an application for payment, showing the amount it considers will be due and how it was calculated, but must do so at least ten days before any due date (cl 7.2). The contractor may prefer to do this, rather than simply wait for a certificate, as it gives it an opportunity to put forward what it thinks the correct figure should be, and why. This information may be very helpful to the contract administrator, but it is not binding; there is no contractual obligation on the contract administrator to accept it, or to provide a detailed rebuttal if it does not agree with the figures put forward (see Figure 6.1).

6.23   The contractor's application is required to show the amount that *will be due*, not what is due at the time of application. As the contract administrator's certificate (see next para.) is required to state the value up to the due date (cl 7.3), not at the date of certification, the contractor should likewise use the due date as its cut-off point when making its assessment of predicted value.

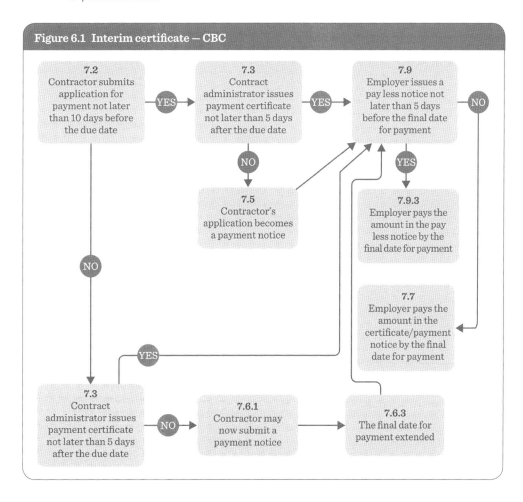

Figure 6.1 Interim certificate — CBC

6.24   Clause 7.3 states:

> No later than 5 days after the Due Date for interim payment, the Architect/Contract Administrator shall issue the Parties with a Payment Certificate showing the amount due and the basis on which it was calculated (even if the amount is zero or negative).

Clause 7.3 then lists what should be covered in the calculation of the amount due. The contract administrator should note that it does not have the power to include or deduct amounts not set out in the clause. In particular, the contract administrator should take care only to include amounts in respect of payments or deductions made by the employer for which it has received written notification, and this should be made clear on the certificate. The certificate should include:

> the total value of work carried out in accordance with the Contract up to the Due Date, plus any additional payments due, minus:

> 7.3.1 the total payment already certified

> 7.3.2 any applicable Retention

> 7.3.3 any applicable Advanced Payment repayment sum

> 7.3.4 any other payments notified to the Architect/Contract Administrator by the Employer

> 7.3.5 any deductions required in accordance with the Contract.

The total value of work carried out

6.25   Note that this should only be work that is in accordance with the contract, and the contract deliberately does not mention materials and goods that are stored on or off site but not yet incorporated in the works, nor does it mention prefabricated off-site items, therefore neither of these should be included in the certificate. This is an important difference to many other standard contracts, which typically require that at least on-site materials etc. are included in the payment certificate.

6.26   The RIBA Building Contracts, unlike some others, do not include 'property vesting' clauses, which provide that such materials and goods will become the property of the client once certified, even if, for example, the contractor has not yet paid its suppliers. Once materials have been fixed to the construction, they will become the property of the client, and cannot be removed by the contractor or a subcontractor. The employer could be at risk, however, where materials have not yet been built in, even where the materials have been certified and paid for. The contractor might not actually own the materials paid for because of a retention of title clause in the contract for the sale of the materials. Under the Sale of Goods Act 1979, sections 16 to 19, property in goods normally passes when the purchaser has possession of them, but a retention of title clause will be effective between a supplier and a contractor even where the contractor has been paid for the goods, provided they have not yet been built in.

6.27   Without property vesting clauses there is, therefore, a danger that unpaid suppliers may return and remove unfixed goods, despite the fact that the client has paid the contractor for them. As there is no requirement in the RIBA Building Contracts to include the value of any of these in interim certificates, normally it would be inadvisable to do so.

6.28   The contract administrator should only certify after having carried out an inspection to a reasonably diligent standard, and should not include any work that appears not to have been properly executed, whether or not it is about to be remedied, and even if the retention is adequate to cover any anticipated remedial work; retention is to cover latent (i.e. hidden) defects, not patent defects (those that are apparent, see *Townsend* v *Stone Toms* and *Sutcliffe* v *Chippendale & Edmondson*). Contract administrators should also note the case of *Dhamija* v *Sunningdale Joineries Ltd*, which stated that a quantity surveyor is not responsible for determining the quality of work executed.

6.29   Where work that has been included in a payment certificate subsequently proves to be defective, the value can be omitted from the next certificate. Under clause 7.3 the contract administrator has the power to issue a 'negative' certificate, although it is arguable whether this would result in an obligation on the contractor to make a repayment. This is because, unlike the final payment provisions (cl 7.15), clause 7.7 does not expressly refer to repayments. This clause derives from the Local Democracy, Economic Development and Construction Act 2009, and on balance it seems unlikely that the contract will be interpreted to allow for negative payments. If such a situation arises, it may be sensible for the employer to seek legal advice, particularly if the amount is significant.

> *Dhamija* v *Sunningdale Joineries Ltd, Lewandowski Willcox Ltd, McBains Cooper Consulting Ltd* [2010] EWHC 2396 (TCC)
>
> An action was brought against the building contractor, the architect and the quantity surveyor (QS) (McBains) for defects in the design and construction of a home. There had been no written or oral contract with the QS, so the terms of its engagement would be those that would be implied. It was held that a QS's contract of retainer would include an implied term that the QS acts with the reasonable skill and care of a QS of ordinary competence and experience when valuing the works properly executed for the purposes of interim certificates, but that the QS would not owe an implied duty to exclude the value of defective works from valuations, however obvious the defects. This was the exclusive responsibility of the architect appointed under the contract. Further, the QS owed no implied duty to report the existence of defects to the architect.

Any additional payments due

6.30   Clause 7.3 also requires the contract administrator to certify 'any additional payments due'. This would cover amounts due to the contractor that are not related to the value of the works, for example the costs and expenses referred to in clause 9.13 (see para. 6.16).

The total payment already certified

6.31   The contract administrator must deduct 'the total payment already certified' (cl 7.3.1). As this deduction is made before retention is deducted, it should be the total gross value of work identified on the previous payment certificate, and before any other deductions are made.

Any applicable retention

6.32   The Explanation of Terms defines 'Retention' as 'a percentage of the amount included in a Payment Certificate that is deducted from a payment in accordance with clause 7'. Under clause 7.4 the retention is set at 5 per cent up until practical completion, and

2.5 per cent during the defects fixing period. The wording does not make it clear exactly what the percentage is deducted from, but its position in the sequence in clause 7.3 suggests it is from the value of work properly carried out, before the other deductions are made (this would tie in with normal practice, e.g. as under JCT contracts). There is no opportunity to specify a different rate to 5 per cent; if a different rate is preferred then the contract would need to be amended.

Any applicable advanced payment repayment sum

6.33   If optional clause A5 has been selected, which allows for the employer to make a payment in advance (see para. 6.20), the parties will have set out a schedule of repayment instalments in item X of the Contract Details. Although clause A5.1 states that the employer should deduct the repayments *from* payment certificates, clause 7.3.3 requires that the deduction is included *within* the certificate. The contract administrator should take care when setting out the calculations to ensure that the contractor is not unintentionally repaid these deductions at a later stage.

Any other payments notified to the contract administrator

6.34   Clause 7.3.4 requires the contract administrator to deduct 'any other payments notified to the Architect/Contract Administrator by the Employer'. This might include payments made to the contractor as a result of a payment notice, in situations where the contract administrator has failed to issue a payment certificate (see para. 6.79).

6.35   However, there should not be any other payments to the contractor of which the contract administrator would be unaware unless 'notified'. Sometimes, on smaller projects, the employer may ask the contractor to, for example, purchase items of equipment not included in the contract documents. This type of direct arrangement should be avoided; all items should be handled through the contract administrator by means of a change to works instruction, as otherwise there are issues as to whether this forms part of the works and is therefore covered by such matters as liability for defects and insurance. However, should such situations inadvertently arise, the clause would enable the contract administrator to deal with them under the main contract, if the parties and the contract administrator agree that it is appropriate.

Any deductions required in accordance with the contract

6.36   There are three key areas of deduction that might be intended in clause 7.3.5:

- Failure to make good defective work – under clauses 5.11 and 10.3.4, the employer may take action if the contractor fails to make good defective work; the contractor will be responsible for the costs and expenses.

- Failure to take out insurance – under clause 6.4, the employer shall 'deduct the cost of the premiums' it incurs when the contractor fails to provide evidence that it has taken out the required insurance (although in DBC this is expressed as a right rather than a duty).

- Liquidated damages – clause 10.1 states that the employer may, on the advice of the contract administrator, deduct liquidated damages.

6.37   In the case of the first deduction, the contract does not state how the costs are to be recovered, but it would be reasonable for the employer to deduct them from payments due. For the other two, however, the contract specifically states that they are to be deductions made by the employer.

6.38   Although there does not appear to be any reason why the deductions cannot be made on the face of the payment certificate, in practice it is more usual, and more sensible, for the employer to make a deduction by means of a pay less notice (see para. 6.51). Whether to make the deduction is entirely a decision for the employer, so if it is made on the certificate, the contract administrator must consult with the employer and ensure that there is a written record of the employer's decision. Alternatively, for clarity the contract administrator could note on the certificate what employer deductions have been made from earlier certified amounts; however, this would simply be a record for information, and would not affect the calculation of the total certified amount.

Interim payments – payment on practical completion

6.39   Optional clause A4.1 states:

> If item W1 is selected in the Contract Details then, subject to clause 7.4, the Employer shall pay the Contractor the Contract Price when the Architect/Contract Administrator has certified Practical Completion.

6.40   Although the clause does not specifically state this, it is clear from the guidance notes that this is intended to replace monthly certification with a single payment at practical completion. As such, it should only be selected when the contract period is very short, and no longer than 45 days (as otherwise the contract must comply with the Housing Grants Act). Unlike in DBC, there is no reference to issuing a certificate to cover this payment, but it is suggested that it would be essential that the contract administrator issues a payment certificate in accordance with clauses 7.3 and 7.4, as discussed above. This payment certificate would, therefore, show the adjusted contract price, with any applicable deductions as listed in clause 7.3, and with 2.5 per cent retention.

Interim payments – milestone payments

6.41   As an alternative to the monthly due dates, the contract allows for payment by milestones (cl A4). If adopted, the parties must define the milestones in detail (item W2 in the Contract Details). The milestones will normally be identifiable points in the completion of the project, for example completion of: (1) foundations and groundworks, (2) ground floor slab and walls up to the damp proof course, (3) external walls, (4) roof, etc.

6.42   Linking payment to milestones introduces an incentive to the contractor to maintain steady progress throughout the project (whereas the threat of liquidated damages applies only to achieving practical completion).

6.43   The exact work that must have been correctly completed for a milestone to be achieved should be set out in detail in item W2. In addition, the parties are required to set out the payment that will be made on achievement of the milestone, either as a value or as a percentage of the contract price. It is suggested that the latter will be more workable, because if the contract price is adjusted due to, for example, variations in the specification, there is no means of adjusting the milestone values, whereas under the percentage system the amount will adjust automatically.

6.44   The corresponding optional clauses for item W2 state:

A4.2   If item W2 is selected in the Contract Details then item P and clauses 7.1, 7.2 and 7.3 on how Due Dates are calculated shall not apply.

A4.3   If item W2 is selected, the Parties will set out the anticipated dates for completion for all milestones and these shall be the Due Dates for payment.

A4.4   When the Architect/Contract Administrator is satisfied that a milestone has been achieved it shall include the payment for that milestone in the Payment Certificate issued for the relevant Due Date.

6.45   Clause A4.3 will need to be applied with care. First, there is no place in item W2 of the Contract Details to insert the 'anticipated dates', therefore these would need to be set out in a separate document. The tense of the clause suggests that this is an obligation on the parties *after* the contract is formed, but (as with dates for normal monthly certification) it would be more sensible to do it before, in order to avoid any misunderstandings.

6.46   Second, if certification is dependent on actual achievement of the milestones (cl A4.4), what is the purpose of agreeing the anticipated dates? In other words, are the certification points: (a) when milestones are actually achieved, or (b) when it is anticipated they will be achieved. This question was raised with the RIBA, which confirmed that (b) is the correct position. If the agreed due date arrives and its associated milestone has not been reached, the payment certificate is nevertheless issued, but showing a value of zero for that milestone. Presumably the contractor will have to wait until the agreed due date for the next milestone to arrive before being paid for the previous one (the RIBA explained that it was necessary to structure the provisions in this way in order for the contract to comply with the Housing Grants Act).

6.47   The milestone payment certificates will therefore state the agreed value or percentage for that stage, once the milestone has been reached. Although the contract does specifically state this, it is suggested that the deductions listed in clauses 7.3.2 to 7.3.5 should also be included, and that retention as set out in clause 7.4 should be applied (clause A4.2 states only that 'clauses 7.1, 7.2 and 7.3 *on how Due Dates are calculated* shall not apply' [emphasis added]; all other aspects of those clauses still apply, as does clause 7.4).

Payment of interim payments

6.48   The employer is required to pay the amount shown on a payment certificate 'on the Final Date for Payment of interim payments', which is 14 days after the due date or any other number of days entered in item P of the Contract Details (cl 7.7). These time periods are quite short: if the contract administrator takes the full five days to issue the certificate, then this will leave the employer with only nine days to pay, which could of course include four weekend days. Clearly, the contract administrator should issue payment certificates promptly, or the parties could consider entering a longer period into the Contract Details.

Contractor's invoice

6.49   Clause 7.8 requires the contractor to issue the employer with a valid VAT invoice, and the employer to pay the invoice promptly. The rules relating to VAT are beyond the scope of this book; if advice is needed the parties should contact HM Revenue and Customs or an appropriate expert. VAT is a matter of law, and any mistakes or attempt to avoid it would be a breach of the law; in this case it would also a breach of contract, allowing the parties to claim against each other for losses due to an infringement.

6.50   It is suggested that the requirement to pay the VAT invoice 'promptly' does not override the time limits as set out in clause 7.7; the employer should pay the amount shown on the payment certificate within 14 days of the due date, regardless of when the VAT invoice is issued. If it is issued shortly after the certificate, the VAT invoice could be paid at the same time, if it arrives later, the VAT will be paid separately.

Pay less notices

6.51   If the employer wishes to pay less than the certified amount, it must issue a 'Pay Less Notice' not later than five days before the final date for payment (cl 7.9.1), and may then pay the amount stated in the notice (cl 7.9.3). Alternatively, the employer may authorise another person to issue the notice on its behalf (cl 7.10). The pay less notice should state the amount the employer considers due, and how it was calculated (cl 7.9.2). The employer has the right to pay less than the certified amount, but only for reasons that can be justified under the terms of the contract. It is suggested that these could include that some of the work covered by the certificate was not in accordance with the contract, or if there was some error in the calculation of the payment certificate, or for any of the matters which might entitle the employer to a deduction, as noted at paragraph 6.36, namely:

- failure to make good defective work;
- failure to take out insurance;
- liquidated damages.

Payments following practical completion

6.52   Following practical completion, the due dates, and therefore certification, continue, but at two-monthly intervals (cl 7.1). With respect to milestone payments, as clause A4.2 states that 'clauses 7.1, 7.2 and 7.3 on how Due Dates are calculated shall not apply' there would be no further certificates until the final payment, unless the parties agree otherwise.

Final contract price and payment

6.53   Most contracts contain provisions for dealing with the final assessment of the contract price. In CBC, clause 7.11.1 requires the contractor to submit its calculation, along with the relevant supporting documentation, to the contract administrator within 90 days of the end of the defects fixing period. In reality, the contractor is likely to submit the material earlier to encourage early certification (see Figure 6.2).

6.54   The contract then requires the contract administrator and the contractor to endeavour to reach agreement on the final amount (cl 7.11.2). If they are unable to reach agreement, or

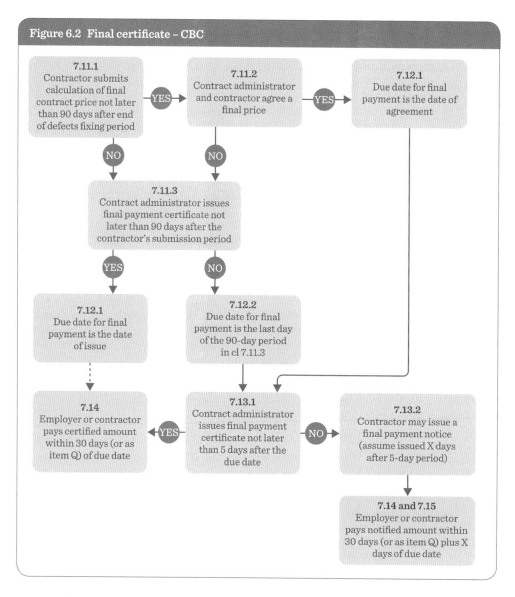

Figure 6.2 Final certificate – CBC

the contractor fails to submit a calculation, the contract administrator should issue a 'final Payment Certificate' not later than '90 days after the Contractor's submission period specified in 7.11.1' (cl 7.11.3). It should be noted that the 90 days runs from the *period*, not from the contractor's submission, so strictly speaking this allows 180 days from the end of the defects fixing period. In practice, however, if the contractor submits the application at an early point in the period, the contract administrator will probably issue the certificate before the full 180 days has run.

6.55   The contract then states: 'The Due Date for final payment shall be the day on which the final Contract Price is agreed under clause 7.11.2 or issued under clause 7.11.3' (cl 7.12.1; presumably this means 'or [the day on which the final Payment Certificate is] issued').

If neither of these occurs, the due date is 'the last day of the 90-day period stated in clause 7.11.3' (cl 7.12.2). Clause 7.12 therefore sets up three possible due dates: at agreement, on issue of the certificate, or 90 days after the contractor's submission period.

6.56   Clause 7.13.1 states: 'No later than 5 days after the Due Date for final payment, the Architect/Contract Administrator shall issue a final Payment Certificate'. This makes sense with respect to the first and third possible due dates (i.e. if the final account is agreed or if no certificate is issued before the end of the period). However, the drafting needs some interpretation as, if followed literally in the case of the second due date, it would result in two final payment certificates; until this is tidied up it is suggested that users assume that clause 7.13.1 does not apply to situations where a payment certificate has already been issued. The employer is required to pay the amount shown on the final payment certificate 'on the Final Date for Payment', which is 30 days after the due date or any other number of days entered in item Q of the Contract Details, subject to any pay less notice (cl 7.15).

Certification and payment – DBC

6.57   The terms setting out payment in DBC are much simpler than those in the CBC. As with CBC, DBC offers the options of monthly certification, a single interim payment following practical completion or milestone payments. However, there is no system of setting up due dates to trigger contractor application, certification and payment deadlines, which makes the procedures more straightforward and easier for the customer to comply with.

Interim payments – monthly certification

6.58   The contract administrator is required to issue certificates at 'the frequency specified in item O of the Contract Details' (cl 7.1). This is either monthly (the default), or once, following practical completion. It is suggested that 'monthly' should be understood as on the same calendar date each month, not four-weekly (as this would reflect the provisions in CBC).

6.59   There is no indication of when the first payment certificate will be issued, and it might be sensible to clarify this in the tender documents. Normal practice is to issue the first certificate one month after the start date. There are, however, no provisions for advance payment in this contract, so in some projects the customer may be prepared to consider an earlier first certificate, for example two weeks after the start date, in order to help cover the contractor's start-up costs.

6.60   Whether to select one payment following practical completion rather than monthly payments will depend upon the length of the project. For small works planned to take less than around six weeks, a single payment might be the most convenient for both parties.

6.61   Under clause 7.1, interim payment certificates should state:

the total value of work carried out in accordance with the Contract to date, plus any additional payments due, minus:

7.1.1 the total payment already certified

7.1.2 any applicable Retention

7.1.3 any notified payments

7.1.4 any deductions required in accordance with the Contract.

6.62   Generally, clause 7.1 follows that in CBC (for a discussion of the valuation of work carried out and the listed deductions, see paras. 6.25–6.38). However, there is one significant difference; clause 7.1 refers to the 'value of work carried out in accordance with the Contract *to date*' [emphasis added], but the 'date' in question is not established; it could be either the date of the valuation or the date of the payment certificate. It is suggested that the former approach should be used.

6.63   Normal practice is that the contract administrator makes its valuation immediately following an inspection (in order to ensure that only work correctly carried out is included). For practical reasons this will almost certainly be a few days before the certificate date. Provided the gap is kept to a minimum, the value shown on the certificate will be close to the value of the work on the day the certificate is issued. However, for the contract administrator to attempt to guess exactly what work will have been carried out between the inspection/ valuation and the certification date and to add that into the value shown could place the customer at risk. As with CBC, it is important that only work correctly carried out is certified, and that unfixed and off-site materials and goods are not included in the valuation (see the discussion of these topics under paras. 6.25–6.27).

6.64   There is one additional minor difference in wording that should be noted: clause 7.1.3 simply states 'any notified payments'. It has been confirmed by the drafters that no difference in meaning was intended to clause 7.3.4 in CBC, discussed at paragraphs 6.34–6.35.

6.65   It should be noted that this contract does not include a provision allowing the contractor to submit an application for payment prior to the certificate date (which is included at cl 7.2 of CBC). There is nothing, however, to prevent the contractor from making such a submission, and in many cases a contractor will submit some form of valuation at regular intervals. In some cases this may be helpful to the contract administrator. However, the contract administrator should remember that this document is of no contractual effect, and should always ensure that the amount shown on a payment certificate is an independent and accurate assessment – the figures shown on the contractor's version should never be simply 'rubber stamped'.

Interim payments – milestone payments

6.66   As with CBC, there is an option for milestone payments (cl A4), which may provide a convenient means of introducing an incentive to progress the project, while also ensuring reasonably regular payment, for projects lasting longer than a few weeks (see paras. 6.41–6.47). Where milestone payments are used, the contract administrator is to issue a payment certificate once a milestone has been reached (cl A4.1; note that DBC does not suffer from the complications of agreeing 'due dates' that arise in CBC), which will state the agreed value or percentage for that stage. Although the contract does not specifically state this, it is suggested that the deductions stated in clauses 7.1.2 to 7.1.4 should also be included.

Payment of interim payments

6.67   Once a payment certificate has been issued, the contractor is required to issue the customer with an invoice. It is important to note that the customer's duty to pay, and conversely the contractor's right to be paid, depends on the customer having received the contractor's invoice – otherwise there is no obligation on the customer to make any payment (there is no equivalent to cl 7.7 in CBC, where the employer's obligation arises on issue of the certificate). The critical date is not the date of the certificate, nor the date of the invoice, but the date of its receipt. The onus is therefore on the contractor to ensure the invoice is issued and delivered promptly (see Figure 6.3).

6.68   Clause 7.3 sets out the procedure for payment as follows:

> 7.3.1 The Contractor shall issue the Customer with an invoice based on the Payment Certificate.

> 7.3.2 The Customer shall pay the invoice within 14 days of receipt.

6.69   Clause 7.7 states:

> 7.7.1 The Contractor shall issue the Customer with a valid VAT invoice for every Payment Certificate.

> 7.7.2 The Customer shall pay the VAT invoice promptly.

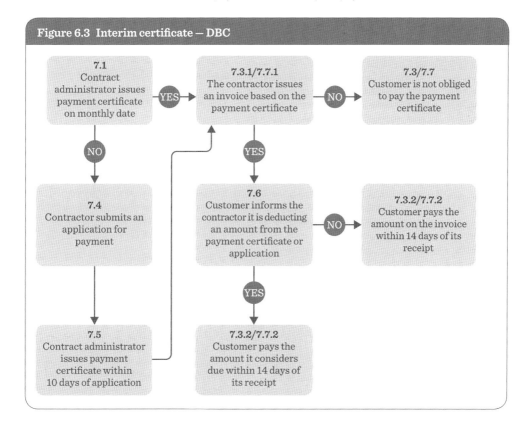

Figure 6.3 Interim certificate – DBC

Both clauses 7.3 and 7.7 refer to the contractor's obligation to issue a payment invoice. These are not presented as alternative clauses, and on enquiry the RIBA has confirmed that it is intentional, and that if the contractor is VAT registered, a VAT invoice should be issued alongside the payment invoice. Nevertheless, it seems unnecessarily complicated, as it would be possible to deal with the matter with one invoice: a clause 7.3 invoice if the contractor is not VAT rated, and a clause 7.7 invoice if it is. It is suggested that the parties consider deleting one of the clauses as appropriate before executing the contract. Issuing two sets of invoices, one without VAT, could cause confusion, and even raise the suspicions of HM Revenue and Customs.

Customer deductions

6.70   Clause 7.6 deals with situations where the customer wishes to pay less than the certified amount. It states:

> If the Customer wants to deduct from a payment an amount not stated in the Payment Certificate or the Contractor's application for payment, it shall inform the Contractor at the time of making the deduction, stating the amount it wishes to deduct and the reason for doing so.

6.71   This differs from CBC in that the customer does not need to notify the contractor five days in advance of making the deduction; the explanation is given at the time of making the deduction, i.e. when making the payment (it is suggested that 'the amount that it wishes to deduct' should more accurately have stated 'the amount it has deducted'). It should also be noted that the customer, not the contract administrator, must inform the contractor, and that there is no equivalent to CBC clause 7.10, whereby the employer can authorise another person to do this on its behalf.

6.72   The clause refers to two situations: deductions from a payment certificate, and deductions from a contractor's application for payment. The second is unnecessary, unlike in CBC, as there is no obligation to make a payment following a contractor's application; as explained above (para. 6.67), the obligation depends on a contractor invoice, which in turn depends on the issue of a payment certificate (which may follow an application). As with CBC, it is suggested that the only deductions that can be made are those allowable under the contract. These could include deductions for work included in the certificate that was not in accordance with the contract, to correct some error in the calculation of the certificate, or for any of the matters that might entitle the customer to a reduction, as noted at paragraph 6.36, namely:

- failure to make good defective work;
- failure to take out insurance;
- liquidated damages.

Payments following practical completion

6.73   DBC does not refer specifically to payments following practical completion but before the final payment, i.e. during the defects fixing period. In the absence of any limiting provision it appears as if payment certificates should continue to be issued at one-monthly intervals (unless milestone payments have been agreed, or a single payment following practical completion). With respect to milestone payments, as with CBC, no further payment certificates would be issued before the final payment, unless the parties agree otherwise.

Final contract price and payment

6.74   In DBC, clause 7.8.1 requires the contractor to submit its calculation, along with the relevant supporting documentation, to the contract administrator within 90 days of the end of the defects fixing period. In reality, the contractor is likely to submit the material earlier to encourage early certification.

6.75   The contract then requires the contract administrator and contractor to endeavour to reach agreement on the final amount (cl 7.8.2). If they are unable to reach agreement, or the contractor fails to submit a calculation, the contract administrator is required to issue a final payment certificate not later than 90 days after the contractor's submission period (cl 7.8.3; apart from the numbering these clauses are identical to those in CBC).

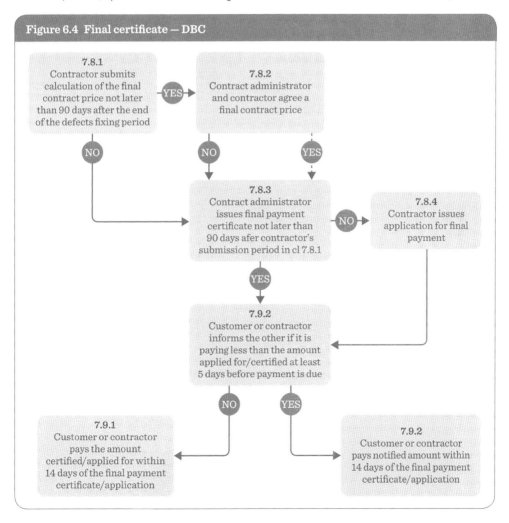

Figure 6.4 Final certificate — DBC

6.76   Somewhat surprisingly, there appears to be no requirement to issue a final payment certificate if the parties do agree on the final amount (i.e. there is no equivalent to cl 7.13.1 in CBC). It is suggested that the contract administrator ought nevertheless to issue a final payment certificate once the amount is agreed (see Figure 6.4).

Conclusiveness

6.77   Unlike in other contracts, it should be noted that no payment certificates are stated to be conclusive evidence that any matters or duties under the contract have been finally discharged, which means that even after the final payment certificate is issued, it is possible for either party to raise a claim for breach of contract. Having said that, it should be noted that if a matter is raised that could have been raised under the currency of the contract, and there is no good reason why the contractual mechanisms were not used to resolve it at that time, then it is unlikely that the claim would be successful.

Non-payment and non-certification

6.78   The contractor is given various remedies should the above payment systems break down, which are different in the two versions of the contract.

Contractor's remedy if no certificate issued

6.79   Under CBC, if no payment certificate is issued, and the contractor had made an application for payment under clause 7.2, then this is said to become a 'Payment Notice' (cl 7.5). The contractor may need to forward the application to the employer, if it has not already done so.

6.80   The employer must pay this amount on the final date for payment for interim certificates (cl 7.5), subject to any pay less notice. If no application had been made, the contractor may now issue a payment notice, showing the amount it considers due and how it was calculated (cl 7.6.1). In this case, the final date for payment is 'extended by the same number of days that it took the Contractor to issue the Payment Notice after the 5-day period in clause 7.3 expired', i.e. after the last date on which the contract administrator ought to have issued the certificate (cl 7.6.3). In either case the employer should notify the contract administrator that payment has been made, so that this can be taken into account in the next certificate. A similar system is set up for the final payment certificate; if the contract administrator fails to issue the certificate, the contractor may submit a final payment notice and the final date for payment is extended (cl 7.13.2 and 7.14).

6.81   Under DBC, which has no earlier right to make an application for interim payment, the equivalent clauses are slightly different. Clauses 7.4 and 7.5 state:

> If the Architect/Contract Administrator has not issued a Payment Certificate in time or the Contractor thinks it is due a payment, the Contractor shall submit an application for payment to the Architect/Contract Administrator stating the amount it considers due and how the amount was calculated.

Within 10 days of receiving an application for payment, the Architect/Contract Administrator shall issue the Parties with a Payment Certificate, even if the amount due to the Contractor is zero or negative.

6.82   Although the intention is clear, there is a possible problem with clause 7.4: what is meant by 'or the Contractor thinks it is due a payment'? As both monthly payments and milestones require a certificate, there could be no reason other than the lack of a certificate that would justify an application. The phrase could be disregarded, except that the contract administrator is obliged to respond to all applications by issuing a payment certificate, even if it disagrees that any certificate is currently due (e.g. because the contractor has applied at the wrong time). It is suggested that the simplest solution in this situation is for the contract administrator to issue a certificate showing a zero amount payable. In addition, the contract administrator should note that there is no obligation to agree with the amount applied for; the contract administrator should always reach its own objective view on what amount should be included.

6.83   If the contractor submits an application under clause 7.4, but the contract administrator does not issue a certificate, the amount applied for does not automatically become payable (unlike in CBC). The contractor's remedy in this situation would be to take the matter to the selected dispute resolution procedure. DBC also provides a remedy if no final payment certificate is issued: the contractor may make an application for final payment showing the final contract price, the amount it considers due and how it was calculated (cl 7.8.4). In this case the customer is obliged to pay the amount shown on the application within 14 days, subject to any notice that it intends to pay less (cl 7.9).

Contractor's remedy if payment not made

Right to interest

6.84   In CBC, if the employer fails to pay an amount that is due by the final date for payment, clause 8.2 makes provision for simple interest to accrue on any unpaid amount (there is no equivalent provision in DBC). The rate of interest can be set by the parties (e.g. they could adopt the rate stipulated in many JCT contracts, i.e. five per cent over the base rate of the Bank of England). If no rate is set in item R, the default is the statutory right to interest under the Late Payment of Commercial Debts (Interest) Act 1998. The interest accrues from the final date for payment until the amount is paid. The provisions apply to the final payment certificate as well as to interim certificates.

6.85   If the employer makes a valid deduction following a pay less notice, it is suggested that interest would not be due on this amount. The clause does not refer to the amount stated on the certificate but to 'any unpaid amount', which would take into account valid deductions.

Right of suspension

6.86   In both versions of the contract the contractor is given a 'right to suspend' under clause 8.1. If the client fails to pay the contractor by the required time limit, the contractor has a right to suspend performance of some or all of its obligations under the contract, which would not only include the carrying out of the work, but could also, for example, extend to any insurance obligations. In addition, under the CBC only, the contractor may request that the employer provides evidence of its ability to pay the contract price, and may suspend its obligations under clause 8 if the evidence is not provided (cl A6).

6.87   This non-payment is stated to be of 'an amount that is due' therefore the contractor may not suspend work if a valid pay less notice has been issued by the employer (CBC) or a notice to pay a lesser amount issued by the customer (DBC). The contractor must have given the client a written seven-day notice of its intention to suspend work and stated the grounds for the suspension, and the default must have continued for a further seven days. The contractor must resume work when the payment is made.

6.88   Under these circumstances the suspension would not give the client the right to terminate the contractor's employment. The suspension will, however, give the contractor the right to a revision of time, and to reasonable expenses and costs arising out of the suspension (cl 8.3). This right is required by the Housing Grants Act, and therefore could be deleted in DBC if the client would prefer.

Termination

6.89   The contractor also has the right to terminate its employment if the client does not pay amounts due (cl 12.3, CBC; cl 12.5, DBC). The contractor must have given a 14-day notice of this intention, which must specify the default and refer to the specific clause. It should be noted that this right is not limited to non-payment of significant amounts, or to persistent non-payment, and so could be exercised for even minor non-payment, provided the notice is given. It would be unlikely, in practice, that the contractor would terminate for minor shortfalls, but it would be unwise for the client to take this risk.

7 Insurance

7.1 Death, injury of people and damage to property are all real possibilities in construction projects and so are risks which any contract needs to address. Almost all contracts, therefore, include provisions to deal with these.

7.2 Three interrelated concepts are usually used to cope with these particular risks: allocation of liability, indemnity and insurance. With respect to liability, if a party is made liable for a risk, it will normally bear the costs of the reasonably foreseeable consequences if that risk materialises. If a party agrees to indemnify another against a risk, it agrees to compensate that party for any losses it suffers should the risk occur. Finally, if a party is required to insure against a loss, the policy it takes out will cover it (and possibly the other party) for any losses.

7.3 The insurance clauses, including the numbering, are the same in the two versions of the RIBA Building Contract, except for the fact that the DBC has an additional optional clause covering insurance-backed guarantees (cl 6.5 and A7).

Liability

7.4 The RIBA Building Contracts allocate liability for insurable risks under clause 6. (These are, of course, not the only risks that each party is liable for; risks are also allocated under, for example, cl 9.1–9.4, which relate to delay.) The contractor's liability is as follows:

6.1.1 loss of or damage to the Works

6.1.2 loss of or damage to products and equipment

6.1.3 death of or bodily harm to any person working for the Contractor arising in connection with the Works during the course of their employment

6.1.4 loss of or damage to property due to the negligence of the Contractor in carrying out the Works

6.1.5 death of or bodily harm to a third person [not an employee of the Contractor] caused by the Works.

7.5 The contractor is therefore liable for the injury or death of any person engaged by it in relation to the works, and that of any third party that is caused by the works. It is also liable for loss or damage to the works, products or equipment, due to any cause, and to property, due to the negligence of the contractor in carrying out the works.

7.6   Under clause 6.2, the client's liability includes:

6.2.1 damage to existing structures and fixtures

6.2.2 damage to neighbouring property caused by the nature of the Works and not due to the negligence of the Contractor in carrying out the Works.

7.7   Although this is generally clear, there is some potential overlap between the allocation of liabilities (i.e. both parties being liable for the same losses). For example, the term 'property' in clause 6.1.4 is not defined, and therefore could include the contractor's property, third party property, the works, and the client's property, including the existing building. However under clause 6.2.1 the client accepts the risk of damage to existing structures, and the clause appears to cover any damage, even if caused by the contractor's negligence. There is therefore a possible overlap between clauses 6.1.4 and 6.2.1.

7.8   It is unlikely that a party would accept liability for damage caused by the other, and the RIBA has confirmed that this was not intended. However, this is the arrangement under many JCT contracts in situations where the employer takes out insurance for the existing structure (e.g. in relation to SBC11 cl 6·2 it has been confirmed by the courts that the contractor is not liable for losses insured by the employer under Option C and caused by a 'Specified Peril', even where the damage is caused by the contractor's own negligence; *Scottish Special Housing Association* v *Wimpey Construction*, *Kruger Tissue* v *Frank Galliers*, *Co-operative Retail* v *Taylor Young Partnerships*, *Scottish and Newcastle* v *G D Construction*). MW11 cl 5·2 contains a similar exclusion of contractor liability for some types of losses caused by its negligence. The purpose of the exclusions is to eliminate a potential overlap, and avoid possible arguments about which insurance policy is meant to cover such losses should they occur. It may therefore be wise for parties to clarify whether the contractor is liable for the losses under 6.1.4 or the client is liable under 6.2.1. (The difficulties experienced in interpreting the less clear wording in earlier versions of MW11 can be seen in *National Trust* v *Haden Young Ltd* and *London Borough of Barking & Dagenham* v *Stamford Asphalt*.)

7.9   There is also potential overlap between clauses 6.1.2 and 6.2.1 (i.e. products and equipment that form part of existing fixtures) and between clauses 6.1.2 and 6.2.2 (i.e. products and equipment that constitute neighbouring property). Again, it is perhaps intended that the 'products and equipment' are those belonging to the contractor and being used as part of the works, but the clause does not make this clear. In all these cases it is likely that, as the clause is clearly intended to distribute liability, a court would construe it in a way that would remove the overlaps as suggested. However, to avoid any arguments it might be sensible for the parties to introduce a clarification before entering into the contract.

7.10   The contractor's liability under clause 6.1 is more widely defined than under the equivalent clauses in JCT contracts (e.g. IC11 cl 6·1 or MW11 cl 5·1), as it does not have the important proviso 'except to the extent that the same is due to any act or neglect of the Employer or of any person for whom the Employer is responsible'. This would mean that under the RIBA Building Contracts, the contractor will be liable even if the injury, etc., was caused by the client, for example by sending other operatives onto the site. Similarly, under JCT contracts the employer is not made liable for the risks in clause 6.2.2 of the RIBA contracts, i.e. for damage to neighbouring property not caused by the negligence of the contractor. A building owner may be liable to its neighbour under common law principles if work it carries out damages the neighbour's property, but this liability is not automatic, and the owner would not normally accept it under the building contract.

Indemnity

7.11   It should be noted that there is no requirement for the contractor to indemnify the client against claims. So, if the client is sued by a party for losses caused by a matter that is the contractor's liability, it may have to settle the claim and then pursue the contractor for compensation. This is in contrast to the typical provisions in JCT contracts (e.g. IC11 cl 6·1 and 6·2 and MW11 cl 5·1 and 5·2), which require indemnification of the employer against claims for injury to or death of persons, or damage to neighbouring property that has been caused by the contractor's negligence.

7.12   In one area, however, the client is given some form of protection. Under the optional contractor design provisions, the contractor is required to 'compensate the Employer/Customer for all claims in respect of the Contractor's designs' (cl A2.6). Although not completely clear, the clause seems to be referring to claims by third parties (not only those by the client, which would go without saying), and therefore constitutes an indemnity. It is quite a widely drafted provision, and would appear to cover damage or losses resulting from the design, even where the contractor has not been negligent, and regardless of whether or not the 'fit for purpose' option has been selected in the Contract Details (see paras. 3.18–3.22).

Insurance

7.13   The purpose of insurance is to ensure that those covered are compensated should the covered risks materialise. In the RIBA Building Contracts, as with most building contracts, the insurance provisions are linked to specific liabilities, and ideally the insurance policies should reflect precisely those liabilities. Even if the insurance is inadequate or non-existent, this would not affect or reduce a party's liability under the contract, or under the law, but in practice it would frequently not have adequate funds to compensate the other party, or affected third parties, for the losses. Any mismatch or lack of clarity in insurance coverage may therefore give rise to arguments, at a time when delays and complications will only exacerbate an already difficult situation.

7.14   Insurance is a complex and specialist subject area. The guidance given here is a brief explanation only, covering some of the key concepts. Which policies will be needed will depend on the particular circumstances of the project; clients are recommended to take specialist advice, and to include any specific requirements in the tender documents.

7.15   DBC requires that 'the Parties shall maintain insurance policies in respect of their liabilities as set out in items L, M and N of the Contract Details' (cl 6.3.1).

7.16   The wording in CBC is slightly different, i.e. 'the Parties shall maintain insurance policies in respect of their liabilities *at the values* set out in items M and N of the Contract Details' (cl 6.3.1). The difference is not significant, as the RIBA has confirmed that the intention behind both these clauses is that the parties are given full flexibility as to their insurance arrangements, and that if no details are set out, there would be no obligation to maintain insurance.

7.17   It is therefore very important that the parties set out full information. The contracts require that the policies taken out by the contractor are to be in joint names (cl 6.3.2), but otherwise no requirements are set out. It is suggested that not only the 'type', 'amount' and 'duration' should be inserted in the Contract Details as required, but that full information about the type of losses to be covered, exclusions, subrogation, etc. should also be given.

7.18   The contractor's liability for injury and death of employees (cl 6.1.3) is met by its employer's liability insurance. This insurance is compulsory under the Employers' Liability (Compulsory Insurance) Act 1969. The legal minimum level of cover for most firms is £5 million, but many insurers will provide a £10 million policy as standard. It is not normal for this type of policy to be taken out in joint names (the JCT contracts do not require it), so the client should check with its advisers whether it is necessary to retain this requirement.

7.19   As noted in the contract guidance notes, the contractor's liability in respect of third parties (death or personal injury and loss or damage to property including consequential loss, covered under cl 6.1.5 and 6.1.4) is met by its public liability policy. Insurers advocate insuring for a minimum of £2 million for any one occurrence, and insurance companies typically offer £5 million as the standard level of insurance.

7.20   As with employer's liability insurance, it is not common to have a public liability policy in joint names, so this should be checked. In addition, a contractor's public liability policy will not usually cover it for damage to neighbouring property, unless this was caused by its own negligence. Under the RIBA Building Contracts, this type of risk is accepted by the client, which is also required to maintain an insurance policy to cover such losses (cl 6.2.2 and 6.3.1). The client will need to check if any existing building insurance it has will cover this risk; if it does not, the client needs to arrange for it to be extended, or take out a new policy. This insurance is usually expensive, and subject to a great many exclusions. The policy needs to be effective at the start of the site operations, when demolition, excavation, etc. are carried out. If the client would prefer the contractor to take out this insurance, the details will need to be given in the tender documents, and the clauses regarding both liability and insurance amended accordingly. (Under SBC11 this type of insurance is an optional provision, taken out by the contractor, and it is not included at all in MW11.)

7.21   With respect to the works, if the project is a new build, the contractor is required take out a joint-names policy to cover any damage to the project (cl 6.1.1 and 6.3.1). With respect to existing buildings, the contractor is required to insure the works, and the client the existing structure (cl 6.1.1, cl 6.2.1 and cl 6.3.1). This is similar to the approach in MW11 clause 5·4B. The arrangements should be discussed with both parties' insurance companies to ensure there are no gaps or overlap. As noted above, if the existing structure is damaged due to a fire caused by the contractor's negligence, is this to be part of the contractor's liability under clause 6.1.4, and, if so, is it to be covered by its public liability insurance (unlikely), or is it intended that the client's existing property insurance will cover such losses?

7.22   One issue that should be considered is that of subrogation, i.e. the right of an insurer to pursue a claim against a third party that caused a loss, in order to recover an amount paid to the insured for that loss. Any joint names policy should make clear that, under the policy, the insurer does not have a right of subrogation to recover any of the monies from either of the named parties. In addition, the client should consider whether the policies should also include a waiver of any rights of subrogation against any subcontractors or required specialists (as they would in JCT contracts).

7.23   It is vitality important that all insurance matters are sorted out before the project starts on site. If there is a gap or an overlap (i.e. both parties insure against the same event or loss) this may cause serious difficulties. The last thing that is needed if a disaster such as a fire should occur is that the contractor or client is unable to honour its liabilities due to lack of funds, and the insurance companies become locked in a dispute and refuse to pay for the essential remedial work. The contractor must provide evidence that adequate insurance has been taken out no later than 10 days before the start date (and any time after on request) – if it is not provided the client may take it out and deduct the costs from payments due (cl 6.4).

Professional indemnity insurance

7.24   Under optional clause A2.7, the contractor is required to ensure that there is adequate professional indemnity insurance for its design responsibilities. A professional indemnity policy insures a firm providing services against losses it suffers due to claims against it for negligence. This type of insurance protects the firm, but consequentially reduces the risk for the client. Should the building suffer defects due to negligent design, and the designer in question has no funds to cover the losses, there would be little purpose in bringing a claim against the designer. However, if an insurance policy is in place, the insurance company will compensate the client. Architects are required by their registration body to have professional indemnity insurance, but it cannot be assumed that a contractor will carry this.

7.25   Two things should be noted about clause A2.7. First, liability may be a 'fitness for purpose' one, if that option is selected in the Contract Details. Second, the contractor is required to 'compensate the Employer/Customer for all claims in respect of the Contractor's design' (cl A2.6). Both these risks might be wider than the risks covered by the contractor's current professional indemnity insurance, so it might be sensible for the client to ask for evidence from the contractor that they have both been satisfactorily dealt with.

8 Termination

8.1   Given the complexity and unpredictability of construction operations, it would be unlikely that a project could proceed to completion without breaches of the contractual terms by one or other of the parties. This is recognised by most construction contracts, which usually include provisions to deal with foreseeable situations. These provisions avoid arguments developing or the need to bring legal proceedings as the parties have agreed in advance machinery for dealing with the breach.

8.2   A clear example of this is the provisions for liquidated damages – the contractor is technically in breach if the project is not completed by the contractual date, but all the consequences and procedures for dealing with this are set out in the contract itself.

8.3   However, some breaches may have such significant consequences that the other party may prefer not to continue with the contract, and for these more serious breaches the contract will contain provisions for terminating the employment of the contractor.

8.4   In any contract, if unforeseen events mean that it becomes impossible for the contractual obligations to be fulfilled, the contract is sometimes said to be 'frustrated' and it may be set aside. In addition, where the behaviour of one party makes it difficult or impossible for the other to carry out its contractual obligations, the injured party might allege prevention of performance and sue either for damages or a *quantum meruit*. This could occur in construction where, for example, the client refuses to allow the contractor access to part of the site. Where it is impossible to expect further performance from a party, the injured party may claim that the contract has been repudiated. Repudiation occurs when one party makes it clear that it no longer intends to be bound by the provisions of the contract. This intention might be expressly stated, or it might be implied by the party's behaviour. In addition to these common law rights, construction contracts normally include express provisions allowing the parties to terminate the contractor's employment should serious situations occur, including but not necessarily limited to those that would amount to frustration or repudiation.

8.5   The idea of terminating the contractor's employment is something that most parties would seek to avoid if at all possible. Dealing with the consequences of termination, and the prospect of having to engage another contractor to complete a half-finished building, are difficult and stressful, and generally parties are better off trying to resolve their differences if at all possible. However, sometimes the situation becomes so difficult that no other option is feasible. In such circumstances, the RIBA Building Contracts, like all other standard contracts, set out reasons and procedures for termination.

Termination by the client

8.6   The client may terminate the contractor's employment under clause 12.2 CBC or clause 12.4 DBC. The reasons listed as giving grounds for termination by the client are covered in clause 12.1 of CBC and clause 12.3 of DBC. These state that the client may terminate the contractor's employment if the contractor:

- abandons the works;
- fails to proceed regularly and diligently;
- consistently fails to comply with instructions;
- is in material breach of the contract.

Abandoning the works

8.7   This would require that the contractor has made no appearance at the site for a significant period of time and has failed to satisfactorily explain why. An absence of a day or two would probably not be significant, but if there is no response to enquiries from the contract administrator, it would be sensible for the contract administrator to issue a notice of termination straight away.

Failing to proceed regularly and diligently

8.8   The phrase 'regularly and diligently' appears in many contracts, and has been the subject of much litigation. The client already has a remedy for slow progress and late completion, in the form of liquidated damages, so a generally poor performance is not normally considered sufficient to justify termination. The phrase means more than simply falling behind any submitted programme, even to such an extent that it is quite clear the project will finish considerably behind time. However, something less than a complete cessation of work on site would be sufficient grounds.

8.9   In the case of *London Borough of Hounslow* v *Twickenham Garden Developments* (1970), for example, the contract administrator's notice was strongly attacked by the defendants. In a more recent case, however, the contract administrator was found negligent because it failed to issue a notice. In *West Faulkner Associates* v *London Borough of Newham* (1992) the court stated:

> Taken together the obligation on the contractor is essentially to proceed continuously, industriously and efficiently with appropriate physical resources so as to progress the works steadily towards completion substantially in accordance with the contract requirements as to time, sequence and quality of work.

8.10   However, although failure to comply with a master programme would not by itself be a breach, it may be some evidence of failure to proceed regularly and diligently. This is where careful records will help to establish a case, and the regular updates to programmes and the records of discussions at progress meetings may be of invaluable assistance.

8.11   In any event, the right to terminate for failure to progress should not be used lightly. In particular, a client that realises the liquidated damages might not compensate it sufficiently should be advised that termination cannot be simply thought of as a convenient alternative. Generally, the failure has to be of reasonable significance, and here the programme would be of considerable help, particularly if it shows the resources to be deployed; if the resources actually deployed are substantially less than planned over a period of weeks, then a default could normally be established.

Consistently failing to comply with instructions

8.12   In JCT contracts, failure by the contractor to comply with just one instruction may be enough to terminate, but in the RIBA Building Contracts, as the plural is used, it appears something more is needed (cl 12.1.3 CBC and 12.3.3 DBC). This may well depend on what is involved. For example, if an instruction relates to health and safety, structural stability, security, compliance with legislation, etc., then failure to comply with repeated instructions relating to the same matter may be enough. Where it is more detailed or cosmetic, failure to comply with numerous instructions regarding a wide variety of matters may be needed.

Material breach of contract

8.13   The concept of a 'material breach' (cl 12.1.4 CBC and 12.3.4 DBC) giving a right to terminate is something that will be unfamiliar to those who normally use JCT contracts. Potentially, it is far wider in coverage than the more specific list of reasons to terminate given in, for example, MW11 (where the defaults include suspension without reasonable cause, failure to proceed regularly and diligently, and failure to comply with the CDM Regulations). However, it is frequently cited as a cause in other contracts, particularly bespoke ones. There are no absolute rules as to what types of breach would be considered 'material'; clearly they would need to be more than simply a trivial or a technical breach, but something less than a repudiatory breach.

8.14   So how serious and significant would a breach need to be to be considered material? Generally, it will depend on the nature of the project, the impact on the client and the circumstances surrounding the breach. The case of *SABIC UK Petrochemicals Ltd* v *Punj Lloyd Ltd* gives some guidance.

> *SABIC UK Petrochemicals Ltd* v *Punj Lloyd Ltd* [2013] EWHC 2916 (TCC), [2013] EWHC 3202 (TCC)
>
> Punj Lloyd was the Indian parent company of an insolvent contractor that had undertaken the construction of a low-density polyethylene plant on the old ICI site at Wilton for SABIC, the employer. SABIC terminated the contract for poor performance and commenced litigation. The court held SABIC's termination justified, although it did not amount to a repudiatory breach. It considered that there were aspects of the contractor's conduct that amounted to deliberate decisions not to comply with all of its contractual obligations, one of which was instructing its subcontractors to demobilise. However, the court's view was that, as at all material times the contractor stated its intention to bring the project to completion, that could not necessarily be equated with a renunciation of its side of the bargain. In that case, the contractor's conduct came close to being repudiatory but 'didn't cross the line'.

8.15   It is suggested that any of the following might constitute a material breach:

- refusal to comply with instructions;
- refusal to remove or correct defective work;

- failure to engage sufficient labour, removal of essential plant from the site;
- failure to comply with statutory obligations, particularly those regarding health and safety;
- any criminal act or evidence of corruption.

Termination by the contractor

8.16   The contractor may terminate its employment under clause 12.4 CBC or clause 12.6 DBC. The reasons listed as giving grounds for termination by the contractor are covered in clause 12.3 of CBC and clause 12.5 of DBC. These state that the contractor may terminate its employment if the client:

- fails to pay the contractor when payment is due;
- is in material breach of the contract.

Failure to pay the contractor when payment is due

8.17   This is not as precise as the equivalent clause in JCT contracts, which states 'does not pay by the final date for payment the amount due … and/or any VAT properly chargeable on that amount' (e.g. IC11, cl 8·9·1·1). In the RIBA Building Contracts there is a certain ambiguity, i.e. 'Fails to pay the contractor [what?] when payment is due'. It is suggested that the 'what' is the amount that has been certified in the case of CBC, or the amount invoiced in the case of DBC, and that the clause does not extend to situations where the contractor is contending that the amount certified/invoiced is less that the amount it considers is 'due'. The 'amount due' should be read in the context of the contract as a whole, i.e. it is as set out in section 7 of the contracts.

Material breach of contract

8.18   As with termination by the client, a breach would need to be more than merely trivial, but not as significant as a full repudiatory breach. It is suggested that any of the following might constitute a material breach by the client:

- refusal to allow access to the site or parts of the site;
- failure to comply with statutory obligations, particularly those regarding health and safety;
- failure to supply information essential for completion of the works;
- employing others to carry out the works;
- seeking to terminate the contractor's employment on grounds not allowed under the contract.

Termination by either party

8.19   Either party may terminate the contractor's employment due to insolvency, bankruptcy or frustration.

Insolvency or bankruptcy

8.20   Clause 12.5 of CBC and clause 12.7 of DBC state:

> If a Party is declared insolvent or bankrupt under any applicable law, the other Party may terminate the Contractor's employment immediately by issuing the insolvent Party with a notice of termination.

8.21   Unlike JCT contracts, no definition of insolvency is given. This may be sensible as the definition under the law may change over time, and this is a matter of statute rather than compliance with contractual provisions. However, as a guide the parties could consult the current definition in an up-to-date JCT contract (e.g. the IC11, cl 8·1) or other up to date text. If there is any doubt about the matter, the client should seek expert advice, as this is an issue that could have serious consequences.

8.22   It should be noted that, unlike in many of the older versions of JCT contracts, termination is not automatic upon insolvency or bankruptcy, and positive action is required by the other party to bring the contractor's employment to an end.

Frustration

8.23   Clause 12.6 of CBC and 12.8 of DBC state:

> Either Party may terminate the Contractor's employment under the Contract if any event not caused by (and not the responsibly of) the Parties has prevented the Contractor carrying out the Works for a continuous period of 60 days.

8.24   The clause is broadly worded, and it is suggested that 'any event not caused by (and not the responsibility of)' should be interpreted in a practical way. For example, the parties have assumed liability for various matters under clauses 6.1 and 6.2 (both contracts) that may not have been caused by them, such as loss of or damage to the works (cl 6.1.1), and similarly have accepted various risks under clauses 9.1–9.4, such as 'damage caused by: fire; lightning; …' (cl 9.1.3). It is quite possible that these events might result in work ceasing for 60 days, and it is suggested that, provided a party did not actually cause the problem, it ought to be able to terminate the contractor's employment should this occur, even though under the contract it accepted liability for or the risk of this event.

Procedure for terminating

8.25   Whatever the grounds for termination, the procedure is the same and follows a two-stage process. First, the contract administrator must issue the contractor with a notice of intention to terminate, referring to clause 12.1 of CBC or clause 12.3 of DBC and stating the reason for the termination. This is an essential first step, as a client that attempted to terminate the contractor's employment without it would be in breach of contract. Although not essential, it may be sensible for the contract administrator to set out exactly what would be needed for the default to be rectified.

8.26   All notices relating to termination are to be sent by recorded delivery as set out in clause 11.12, and are effective from the date of delivery. In addition, as this is a serious step, which could ultimately bring the project to an end, it would be advisable for the contract administrator to discuss it with the client beforehand.

8.27   At the second stage, the client may issue a notice of termination (cl 12.2 CBC and 12.4 DBC). Before the notice can be issued, the contractor must have failed to remedy the default within 14 days of receiving the initial notice of intention to terminate.

Consequences of termination

8.28   The consequences of termination are set out in clauses 12.7–12.10 (CBC) and clauses 12.9–12.12 (DBC). The clauses are primarily concerned with payment.

Payment

8.29   The contracts explain how the balance due is to be calculated in situations where the client has terminated the contractor's employment, and in situations where the contractor initiated the termination (cl 12.7, CBC; cl 12.9, DBC). They also establish *when* the balance is to become due (cl 12.8 and cl 12.9, CBC; cl 12.10 and cl 12.11, DBC).

8.30   Unfortunately, in the case of CBC these clauses are not as clear as they might be. Clause 12.8 says: 'Subject to clause 12.9.3, the balance shall be due for payment within 30 days of either …'. A little later, clause 12.9.1 states: 'If the Employer terminates the Contractor's employment under the Contract, no further payment shall become due to the Contractor', and clause 12.9.3 states: 'Payment shall only become due after the Works are completed or, if the Works have been discontinued, 180 days after termination'.

8.31   Clauses 12.9.1 and 12.9.3 appear to contradict each other; the first states that payment is not due, but the second states that payment shall become due, but only after a specific period.

8.32   In unpicking these clauses, first it has to be assumed that all parts of clause 12.9 deal with the situation when the employer has terminated (although the clause is not structured that way). Second, clause 12.9.1 is probably intended to mean 'shall not become due immediately' or perhaps 'shall not become due within 30 days'. Third, if this analysis is correct, then the 30-day limit in clause 12.8 will only apply in situations where the contractor has terminated its employment, and is not relevant to situations where the employer has. The sequence is illustrated in Figure 8.1.

8.33   It should be noted that clause 12.9.2 contains an important protection for the employer, i.e. that any amount that may have become due by the time of the termination does not need to be paid in situations where the employer has issued a pay less notice, or if the contractor has become insolvent before the final date for payment (this reflects the position as set out by the House of Lords in *Melville Dundas* v *George Wimpey*).

8.34   In the case of DBC the timing is more straightforward. If the contractor terminates its employment, the balance is due within 14 days of the contractor submitting an application for payment, or of the contract administrator issuing a payment certificate, whichever occurs first (cl 12.10). If the customer terminates, then 'clause 12.10 shall not apply until the Works are completed or, if the Works have been discontinued, until 180 days after

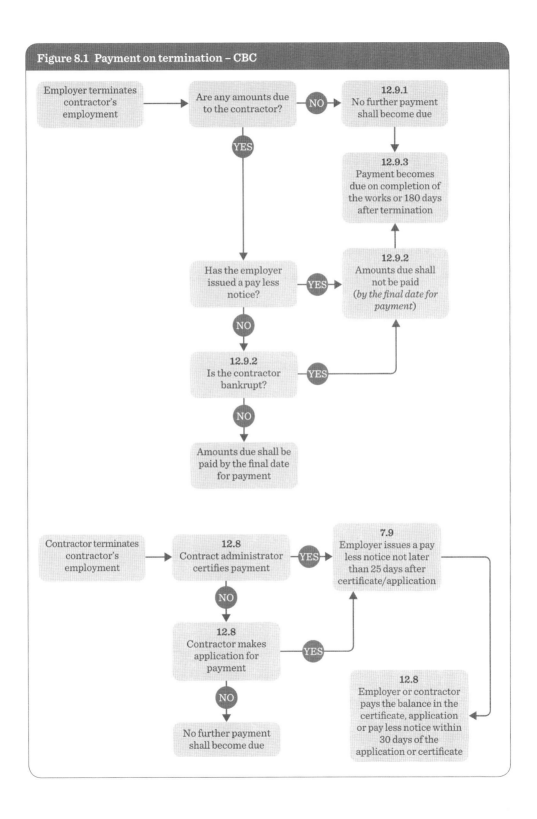

Figure 8.1 Payment on termination – CBC

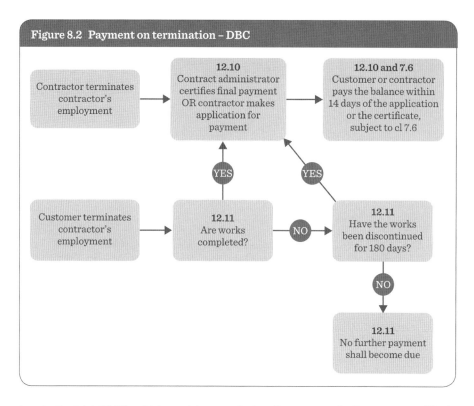

Figure 8.2 Payment on termination – DBC

termination' (cl 12.11), which would mean that neither an application nor a certificate should be issued until after one of these conditions is met. The sequence is illustrated in Figure 8.2.

Access to the site and security

8.35   If the contractor's employment under the contract is terminated for any reason, clause 12.10 (CBC) and clause 12.12 (DBC) state that the contractor shall:

- lose its right to access the site;
- remove all its materials and equipment within a reasonable time;
- no longer be responsible for the security of the site.

8.36   This is a sensible clause, which makes the position clear to the parties should termination occur. The client should note that it will immediately become responsible for the site, and will probably need to take steps to ensure that it is secure and complies with any health and safety or other statutory requirements. The client will also need to allow the contractor access to remove its materials and equipment; the contractor should give reasonable notice of when it intends to do this so that arrangements can be made.

9 Dispute handling and resolution

9.1   At the start of a project the idea that the parties may fall into dispute may seem only a remote possibility and therefore not worth considering in detail. However, the reality is that once a dispute has arisen it will not be possible to sensibly agree a way forward to its resolution. It is therefore very important that care is taken at the outset to select dispute resolution options that are appropriate and with which both parties are happy.

9.2   Ideally, if a dispute breaks out it should first be handled informally, through negotiation, correspondence or, perhaps, at a specially called meeting. In fact, both of the RIBA Building Contracts require that the parties meet and negotiate to resolve disagreements (cl 13.1). The advance warning notices and related meetings will also help to avert or resolve many matters. Embarking on more formal dispute resolution methods is not something to be undertaken lightly; they will all involve the parties in additional costs and time (although to varying degrees) and may be very stressful. In addition, with the exception of mediation, they are likely to result in a poor relationship between the parties for the remainder of the project.

9.3   However, in some cases the need for a more formal method of dispute resolution will be unavoidable. For such situations, the RIBA Building Contracts offer three options: adjudication, mediation and arbitration.

9.4   In DBC, all three methods can be selected, or none, or any combination; in CBC the adjudication option is preselected (see paras. 2.24–2.25), but otherwise the same applies. The parties are invited to name the arbitrator, adjudicator or mediator of their choice, but if none is named, the parties may apply to a body named to make the appointment. If no body is named, the default body is the RIBA.

9.5   A brief outline of each method is set out below, in order to highlight the key differences between the procedures. More information can be found in the many texts on these topics, including the *RIBA Good Practice Guides* on mediation, arbitration and adjudication listed at the end of this Guide.

Mediation

9.6   Mediation is a voluntary process whereby the parties are helped by a professional mediator to resolve the dispute in a way that they are both comfortable with. The mediator has no authority to impose a solution, and it may be that the dispute is not resolved. However, as it is less adversarial than other methods, even if a solution is not reached the mediation may have helped to pave the way for further discussions. It is also less expensive than other methods, particularly if the mediation is limited to one day. The parties should bear in mind that they will nevertheless need to pay their own expenses, the mediator's fee and the cost of the venue.

9.7   If this option is selected, the mediation would normally happen before any other steps are taken. In CBC it is not stated to be a 'condition precedent', so if it is selected, there would be nothing to prevent either party initiating adjudication, arbitration or court proceedings at the same time. DBC states: 'If mediation is selected then all disputes shall first be referred to the Mediator identified in item P' (cl 13.2). If a party ignores this clause and proceeds straight to, for example, adjudication, then the other party ought to draw the adjudicator's attention to clause 13.2, as it is possible that the adjudicator may decide that he or she has no jurisdiction to hear the dispute until the mediation has taken place. However, case law on this point is limited (e.g. see *Emirates Trading Agency LLC* v *Prime Mineral Exports Private Ltd*), and it may be sensible for the party challenging the adjudication to take legal advice.

9.8   The RIBA offers a mediation scheme, whereby the mediator will be appointed to work with the parties, and is normally paid on a daily rate.[1] For a small project, the mediation may last for a day or even less, but for large, complex disputes it could take several days. There are no set rules for mediation, and each mediator will have various techniques that they may use, for example meeting the parties separately, taking issues and suggestions back and forward between the two, and/or bringing them together for a chaired discussion. All proceedings are confidential and on a 'without prejudice' basis, so that offers made cannot be raised in future dispute proceedings, and nothing disclosed to the mediator will be disclosed to the other party unless a party permits it.

9.9   At the end of the mediation, if an agreement is reached, the parties will sign a binding agreement, which is enforceable in the same way that any contract between the parties would be. If the parties are still in deadlock they can ask the mediator to propose a solution, but of course they do not have to accept it.

Adjudication

9.10   Adjudication is a statutory right of any party to a construction contract that falls under the definition in the Housing Grants Act (see para. 1.14). It is a procedure by which the dispute may be referred by either party to an adjudicator, who must reach a decision within 28 days. In construction industry terms, this is a relatively short period, although it can be extended by 14 days by the referring party, and further by agreement.

9.11   Adjudication has advantages and disadvantages; the key advantage being that the matter is sorted out quickly, which limits the amount of time and resources that the parties can spend on it. The disadvantage is that the whole process can seem extremely rushed, particularly from the client's perspective. Usually, the dispute is initiated by the contractor, who may have spent a considerable period in advance preparing for the process, but the client may be required to respond very quickly to what can be a lengthy claim. The decision may seem rather 'rough justice' as there is little time for the responding party to develop and put forward careful arguments, or to commission technical reports and collect other relevant evidence. On the plus side, the decision, although it must be complied with, can be challenged in the sense that the dispute can later be raised in arbitration or in court (see paras. 9.18–9.19).

---

[1] Details can be downloaded at www.architecture.com/Files/RIBAProfessionalServices/ProfessionalConduct/DisputeResolution/Mediation/RIBAMediationScheme.pdf

The adjudication process

9.12   Neither version of the RIBA Building Contract sets out a procedure for adjudication. Instead, CBC refers to secondary legislation, the Scheme for Construction Contracts (the Scheme,[2] see para. 1.15), whereas DBC refers to the RIBA Adjudication Scheme for Consumer Contracts (the RIBA Scheme). The most significant difference between these is that the latter is a shorter procedure, and the adjudicator's fees are capped.

9.13   Under the statutory Scheme, the party wishing to refer a dispute to adjudication must first give notice to the other party identifying briefly the dispute or difference, giving details of where and when it has arisen and setting out the nature of the redress sought. If no adjudicator is named, the parties may either agree an adjudicator or either party may apply to the nominating body identified in the Contract Details (the default is the RIBA). The adjudicator will normally then send terms of appointment to the parties. In the case of the RIBA Scheme, the party wishing to have the dispute resolved by adjudication must apply to the RIBA.

9.14   Under the statutory Scheme, the referring party must refer the dispute to the selected adjudicator within seven days of the date of the notice (Scheme para. 7(1)). The referral will normally include particulars of the dispute, and must include a copy of, or relevant extracts from, the contract, and any material the party wishes the adjudicator to consider (para. 7(2)). A copy of the referral must be sent to the other party, and the adjudicator must inform all parties of the date it was received (para. 7(3)). The adjudicator will then set out the procedure to be followed. A preliminary meeting may be held to discuss this, otherwise the adjudicator will send the procedure and timetable to both parties. The party that did not initiate the adjudication (the responding party) will be required to respond by a stipulated deadline. The adjudicator may hold a short hearing and/or may visit the site. Occasionally, it may be possible to carry out the whole process by correspondence (often termed 'documents only'). The adjudicator must reach a decision within 28 days of the referral, unless extended by agreement.

9.15   The RIBA Scheme does not set out any particular procedure. It is left entirely to the discretion of the adjudicator to direct the parties as to whether and when referral documents and a response are to be submitted, but in most cases the adjudicator will ask for information from each side and may, as above, decide to hold a meeting. In this case the decision must be reached within 21 days of the adjudicator being appointed.

9.16   The adjudicator is required to act impartially (Scheme paragraph 12 (a); RIBA Scheme paragraph 17). The statutory Scheme states that the adjudicator is not liable for anything done or omitted when acting properly as an adjudicator (para. 26).

9.17   Under the statutory Scheme the parties must meet their own costs of the adjudication, unless they have agreed that the adjudicator shall have the power to award costs, which they may only do after the dispute has arisen. The adjudicator, however, is entitled to charge fees and expenses (subject to any agreement to the contrary) and may apportion those fees between the parties. The parties are jointly and severally liable to the adjudicator for any sum that remains outstanding following the adjudicator's determination. This means that in the event of default by one party, the other party becomes liable to the adjudicator for the outstanding amount. In the case of the RIBA Scheme the parties must

---

2   See http://www.legislation.gov.uk/uksi/1998/649/contents/made

meet their own costs. The adjudicator's fees are limited to £100 per hour (exclusive of VAT) up to a maximum of ten hours, and the parties are jointly and severally liable for the full amount.

Challenging an adjudicator's decision

9.18   The Scheme states that the adjudicator's decision will be final and binding on the parties 'until the dispute is finally determined by legal proceedings, by arbitration (if the contract provides for arbitration or the parties otherwise agree to arbitration) or by agreement between the parties' (para. 23(2)). The effect of this is that if either party is dissatisfied with the decision, it may raise the dispute again in arbitration or litigation as indicated in the Contract Details, or it may negotiate a fresh agreement with the other party. In all cases, however, the parties remain bound by the decision and must comply with it until the final outcome is determined.

9.19   The RIBA Scheme states that 'The customer and the contractor must follow the adjudicator's decision as part of their obligations under the building contract, unless and until either party obtains a court judgment about the dispute which is different from the decision of the adjudicator' (para. 22). There is no reference to final determination by arbitration, possibly because the RIBA Scheme was originally developed to cater for the JCT Building Contracts for a Home/Owner Occupier, which do not include arbitration as a dispute resolution option. Users of a RIBA Domestic Contract that intend to select arbitration as the final forum may wish to include an appropriate amendment to paragraph 22 under item P.

9.20   If either party refuses to comply with the decision, the other may seek to enforce it through the courts. Generally, actions regarding adjudicators' decisions have been dealt with promptly by the courts and the recalcitrant party has been required to comply. Paragraph 22A of the Scheme allows the adjudicator to correct clerical or typographical errors in the decision, within five days of it being issued, either on the adjudicator's own initiative or because the parties have requested it, but this would not extend to reconsidering the substance of the dispute.

Arbitration

9.21   Arbitration is essentially a private alternative to court proceedings. If arbitration is not selected, the default process for the ultimate resolution of disputes will be the courts (cl 13.5). If it is selected, either party has the right to require that any dispute is taken to arbitration (cl 13.4). If one of the parties nevertheless initiates court proceedings, the other can apply for a stay, and the courts would normally freeze the proceedings to allow the arbitration to take place.

9.22   The process is supported by statute (the Arbitration Act 1996), and a court would enforce the decision ('award') of an arbitrator in the same way that it would enforce its own judgments. There are very limited rights to challenge the enforcement of an award, the grounds of which are confined to matters such as lack of jurisdiction of the arbitrator or bias, in some circumstances a party may appeal the award on a point of law, but even this right can be excluded if the parties agree.

9.23   Arbitration is a private process, which is often an attractive feature to clients and their consultants. It is also very flexible, as the parties can agree the timetable and venue; failing agreement, the arbitrator has authority to direct these matters. The form of proceedings can vary hugely, but they are often relatively long and formal, and consequently expensive in comparison with mediation and adjudication. Various sets of rules exist that the parties can adopt, such as the Construction Industry Model Arbitration Rules (CIMAR).[3] These include rules for a 'documents only' arbitration (i.e. with no hearing, see Rule 8), for a short hearing (Rule 7) and for a full procedure (Rule 9).

9.24   Under the Arbitration Act the arbitrator has the power to award costs, unless the parties agree otherwise. Where the arbitrator has the power to award costs, this will normally be done on a judicial basis, i.e. the loser will pay the winner's costs (CIMAR Rule 13.1). The arbitrator will be entitled to charge fees and expenses and will apportion those fees between the parties on the same basis. The parties are jointly and severally liable to the arbitrator for fees and expenses incurred.

9.25   As the costs in an arbitration are often significant (sometimes amounting to more than the actual amount claimed), the issue of who pays them is of considerable concern to the parties. In an effort to reduce their liability for costs, the party which is being claimed against may make an offer to settle. If this is done correctly, and the other party refuses the offer but ultimately is awarded less than the sum offered, it will not be able to recover any of its costs from the date the offer was refused.

---

[3]   These can be downloaded at www.jctltd.co.uk/docs/JCT_CIMAR%202011.pdf

References

Publications

Birkby, G, Ponte, A and Alderson, F. *Good Practice Guide: Extensions of Time*, RIBA Publishing, London (2008).

Construction Industry Council. *Construction (Design and Management) Regulations 2015*. Risk Management Briefing. CIC, London (2015).

Coombes Davies, M. *Good Practice Guide: Adjudication*, RIBA Publishing, London (2011).

Coombes Davies, M. *Good Practice Guide: Arbitration*, RIBA Publishing, London (2011).

Eggleston, B. *Liquidated Damages and Extensions of Time: In Construction Contracts*, 3rd edition, Sweet & Maxwell, London (2009).

Finch, R. *NBS Guide to Tendering: For Construction Projects*. RIBA Publishing, London (2011).

Furst, S. and Ramsey, V. (eds.). *Keating on Construction Contracts*, 9th edition, Sweet & Maxwell, London (2012).

Grossman, A. *Good Practice Guide: Mediation*, RIBA Publishing, London (2009).

Health and Safety Executive. *Managing Health and Safety in Construction*, HSE Books (2015).

JCT. *Tendering Practice Note 2012*. Sweet & Maxwell, London (2012).

Ostime, N. *RIBA Job Book*, 9th edition, RIBA Publishing, London (2013).

RICS and Davis Langdon, *Contracts in Use: A Survey of Building Contracts in Use During 2010*, RICS, London (2012).

Cases

Alfred McAlpine Capital Projects Ltd v *Tilebox Ltd* [2005] BLR 271	2.18
City Inn Ltd v *Shepherd Construction Ltd* [2008] CILL 2537 Outer House Court of Session	4.40
Co-operative Retail Services Limited v *Taylor Young Partnerships* [2002] BLR 272	7.8
Dhamija v *Sunningdale Joineries Ltd, Lewandowski Willcox Ltd, McBains Cooper Consulting Ltd* [2010] EWHC 2396 (TCC)	6.29
Emirates Trading Agency LLC v *Prime Mineral Exports Private Ltd* [2014] EWHC	9.7
Gloucestershire County Council v *Richardson* [1969] 1 AC 480 HL	3.29
H W Nevill (Sunblest) Ltd v *William Press & Son Ltd* (1981) 20 BLR 78	4.3, 5.34
Hamid v *Francis Bradshaw Partnership* [2013] EWCA Civ 470	2.37
Henry Boot Construction (UK) Ltd v *Malmaison Hotel (Manchester) Ltd* (1999) 70 Con LR 32 (TCC)	4.40
Impresa Castelli SpA v *Cola Holdings Ltd* (2002) CLJ 45	4.3, 5.44
Jameson v *Simon* (1899) 1 F (Ct of Session) 1211	5.17
Kruger Tissue v *Frank Galliers* (1998) 57 Con LR 1	7.8
London Borough of Barking & Dagenham v *Stamford Asphalt Co. Ltd* (1997) 82 BLR 25 (CA)	7.8
London Borough of Hounslow v *Twickenham Gardens Development* (1970) 78 BLR 89	4.3, 8.9
McGlinn v *Waltham Contractors Ltd* [2007] 111 Con LR 1	5.17
Melville Dundas Ltd v *George Wimpey UK Ltd* [2007] 1 WLR 1136 (HL)	8.33
Mul v *Hutton Construction Limited* [2014] EWHC 1797 (TCC)	6.12
National Museums and Galleries on Merseyside (Trustees of) v *AEW Architects and Designers Ltd* [2013] EWHC 2403 (TCC)	3.13, 5.10

The National Trust for Places of Historic Interest and Natural Beauty v *Haden Young Ltd*
    (1994) 72 BLR 1 (CA)   7.8
Peak Construction (Liverpool) Ltd v *McKinney Foundations Ltd* (1970) 1 BLR 111 (CA)   4.24
Pearce and High Ltd v *John P Baxter and Mrs A S Baxter* [1999] BLR 101 (CA)   5.47
The Queen in Rights of Canada v *Walter Cabbott Construction Ltd* (1975) 21 BLR 42   4.3
RWE Npower Renewables v *JN Bentley Ltd* [2014] EWCA Civ 150   2.16
SABIC UK Petrochemicals Ltd v *Punj Lloyd Ltd* [2013] EWHC 2916 (TCC), [2013]
    EWHC 3202 (TCC)   8.14
Scottish and Newcastle plc v *G D Construction* [2003] BLR 131   7.8
Scottish Special Housing Association v *Wimpey Construction* (1986) 34 BLR 1   7.8
Skanska Construction (Regions) Ltd v *Anglo-Amsterdam Corporation Ltd* (2002)
    84 Con LR 100   5.44
Sutcliffe v *Chippendale & Edmondson* (1971) 18 BLR 149   6.28
Tameside Metropolitan Borough Council v *Barlow Securities Group Services*
    Limited [2001] BLR 113   2.36
Temloc Ltd v *Errill Properties Ltd* (1987) 39 BLR 30 (CA)   2.18, 2.19
Townsend v *Stone Toms & Partners* (1984) 27 BLR 26 (CA)   6.28
Walter Lilly & Co Ltd v *Giles Mackay & DMW Ltd* [2012] EWHC 649 (TCC)   3.14, 4.39
Wells v *Army & Navy Co-operative Society Ltd* (1902) 86 LT 764   5.10
West Faulkner Associates v *London Borough of Newham* (1992) 61 BLR 81   8.9
Westfields v *Lewis* [2013] EWHC 376 (TCC) 27 February 2013   1.16

Legislation

Statutes

Arbitration Act 1996   9.22
Consumer Right Act 2015   1.28
Contracts (Rights of Third Parties) Act 1999   2.30
Employers' Liability (Compulsory Insurance) Act 1969   7.18
Freedom of Information Act 2000   Table 3.3
Housing Grants, Construction and Regeneration Act 1996, as amended by
    Part 8 of the Local Democracy, Economic Development and
    Construction Act 2009 (the Housing Grants Act)   1.3, 1.12, 1.13, 1.14–1.19,
    1.20, 1.22, 2.24, 6.19, 6.21,
    6.40, 6.46, 9.10
Late Payment of Commercial Debts (Interest) Act 1998   1.22, 6.84
Limitation Act 1980   2.12
Local Democracy, Economic Development and Construction Act 2009   1.3, 6.29
Party Wall etc. Act 1996   2.20
Sale of Goods Act 1979   6.26

Statutory instruments

Construction (Design and Management) Regulations 2015 (CDM Regulations)   5.6, 8.13
Consumer Contracts (Information, Cancellation and Additional Charges)
    Regulations 2013   1.27
Scheme for Construction Contracts (England and Wales) Regulations 1998   1.15, 9.12
Unfair Terms in Consumer Contracts Regulations 1999   1.5, 1.19, 1.25

Further Reading

Chapter 1

Lupton, S. *et al. Which contract?: choosing the appropriate building contract*, 5th editionn, RIBA Publishing, London (2012).
Phillips, R. *A Short Guide to Consumer Rights in Construction Contracts*, RIBA Publications Ltd, London (2013).

Chapter 2

Aeberli, P. *Focus on Construction Contract Formation.* RIBA Publishing, London (2003).

Chapter 4

Birkby, G., Ponte, A. and Alderson, F. *Good Practice Guide: Extensions of Time*, RIBA Publishing, London (2008).
Eggleston, B. *Liquidated Damages and Extensions of Time: In Construction Contracts*, 3rd edition Sweet & Maxwell, London (2009).
Haidar, A. and Barnes, P. T. *Delay and Disruption Claims in Construction: A Practical Approach*, RIBA Publications Ltd, London (2014).

Chapter 5

Bussey, P. *CDM 2015: A Practical Guide for Architects and Designers*, RIBA Publishing, London (2015).

Chapter 6

Haidar, A. and Barnes, P. T. *Delay and Disruption Claims in Construction: A Practical Approach*, RIBA Publications Ltd, London (2014).
Whitfield, J. *RIBA Good Practice Guide: Assessing Loss and Expense*, RIBA Publications Ltd, London (2013).

Chapter 7

Hogarth, R., Anderson, A. and Goldring, S. (eds) *Insurance Law for the Construction Industry*, 2nd edition, OUP, Oxford (2013).

Chapter 9

Coombes Davies, M. *Good Practice Guide: Adjudication*, RIBA Publishing, London (2011).
Coombes Davies, M. *Good Practice Guide: Arbitration*, RIBA Publishing, London (2011).
Grossman, A. *Good Practice Guide: Mediation*, RIBA Publishing, London (2009).
Rawley, D. QC, Martinez, M., Williams, K. and Land, P. *Construction Adjudication and payments*, OUP, Oxford (2013).

Clause Index *by paragraph number*

CBC cl	DBC cl	paragraph
Agreement		2.12, 2.37
Items		
C	C	2.13, 4.2
D	D	2.14
E	E	2.14
F	F	2.15–2.16
G		3.43
H	G	2.13
I	H	2.13
J	I	2.17–2.19
K	J	5.46
L	K	2.2, 3.34
M		7.16
	L	7.15
N		7.16
	M	7.15
	N	2.21, 7.15
O	O	2.22–2.23, 6.6, 6.58
P		6.21, 6.48
Q		6.56
R		6.84
S		2.24–2.25, 9.19
	P	2.24–2.25
T	Q	4.8, 4.10, 4.11
U	R	2.26–2.27, 3.12, 3.19,5.12
V	S	2.17, 2.19, 4.7
W1		6.39
W2		6.41, 6.45
	T	6.66
X		6.20, 6.33
Y		1.22, 6.86
Z	U	3.27
	V	3.37
	W	2.21
AA		2.28–2.30
BB		1.3

CBC cl	DBC cl	paragraph
CC	X	3.44
DD	Y	2.31
Clauses		
1.1	1.1	3.9
1.1.1	1.1.1	3.9, 4.4
1.1.2	1.1.2	3.9, 3.33, 4.4, 4.16, 4.2
1.1.3	1.1.3	3.9, 3.34
1.2	1.2	3.10
1.3	1.3	5.4
1.4	1.4	4.2
1.5	1.5	2.13
1.6	1.6	3.23
1.7	1.7	3.23, 3.26, 3.29
2.1	2.1	4.3, 5.38
2.2	2.2	2.14
2.3	2.3	2.14
2.4	2.4	4.6
3.1	3.1	3.40
3.2	3.2	4.22
3.3.1	3.3.1	4.23
3.3.2	3.3.2	4.35, 4.37
3.4	3.4	6.9
3.5	3.5	6.10
3.6		3.43
4.3	4.3	3.33, 5.6–5.8
5.1	5.1	3.4–3.7, 5.10
5.2.1		2.38
5.2.2	5.2.2	2.38
5.3.1	5.3.1	5.16
5.3.2	5.3.2	5.16
5.3.3	5.3.3	5.19
5.4.1	5.4.1	2.45
5.4.3	5.4.3	5.19–5.20
5.4.4	5.4.4	2.44–2.45, 4.17
5.4.7	5.4.7	5.13
5.4.8	5.4.8	4.17, 4.42, 5.22–5.23

CBC cl	DBC cl	paragraph	CBC cl	DBC cl	paragraph
5.5	5.6	5.24		7.8.1	6.74–6.75
5.6	5.7	5.24–5.28	7.11.2		6.54
5.7	5.5	5.19, 6.11		7.8.2	6.74–6.75
5.8	5.5.3	5.20, 6.12	7.11.3		6.55
5.9		5.29		7.8.4	6.83
5.9.2		3.33	7.12		6.55
5.10		5.29	7.13		6.56
5.11		5.30, 6.36		7.8	6.83
	5.8	5.30		7.9	6.83
5.12		2.40, 2.45, 4.27, 5.31, 6.7, 6.13–6.17	7.14		6.80
	5.9	2.45, 4.27, 5.31, 6.7, 6.13–6.17	7.15		6.29, 6.56
			8.1	8.1	6.86–6.88
5.13		2.40, 5.31, 6.13	8.2		6.84
	5.10	5.31, 6.13	8.3	8.2	6.88
5.13.1	5.10.1	4.29	9.1	9.1	4.26
5.13.2	5.10.2	4.37, 6.13–6.15	9.2	9.2	4.26
5.14	5.11	2.44	9.3	9.3	4.26
5.15	5.12	2.44	9.4	9.4	4.26
6.1	6.1	7.4, 7.7–7.10, 7.18–7.19, 7.21	9.5	9.5	4.26
			9.6	9.6	4.26
6.2	6.2	7.6, 7.7–7.10, 7.20–7.21	9.8	9.8	4.26
6.3	6.3	7.15–7.17, 7.20, 7.21	9.9	9.9	2.45, 4.25–4.33
6.4	6.4	6.36, 7.23	9.10	9.10	4.27–4.33
	6.5	2.21, 7.3	9.11.1		4.29
7.1		6.21, 6.52	9.11.2		4.28–4.31, 4.37, 4.42
7.2		6.22, 6.65, 6.79	9.12	9.12	4.29
7.3		6.23–6.24, 6.29–6.34, 6.36, 6.40, 6.47	9.13	9.13	6.7, 6.16–6.17, 6.3
			9.13.1	9.13.1	6.16
	7.1	6.58, 6.61–6.62, 6.64–6.65	9.13.2	9.13.2	6.16
			9.14	9.14	6.17
	7.3	6.67–6.69	9.15	9.15	5.32, 5.35
7.4		6.32, 6.40, 6.47	9.16	9.16	5.36, 5.37
7.5		6.79–6.80	9.17	9.17	5.40
7.6		6.80	9.18	9.18	5.41
	7.4	6.81–6.83	10.1	10.1	5.45, 6.36
	7.5	6.81	10.2	10.2	5.46
	7.6	6.70–6.72	10.3.1		5.38, 5.46
7.7		6.29, 6.48, 6.50, 6.80	10.3.2		5.48
	7.7	6.67–6.69	10.3.3		5.47, 5.48
7.8		6.49	10.3.4		5.47, 6.36
7.9		6.51	11.1	11.1	2.42
7.10		6.51	11.2	11.2	2.43
7.11.1		6.53, 6.54, 6.55	11.5	11.5	2.14, 3.4
			11.6	11.6	2.30

CBC cl	DBC cl	paragraph
11.9	11.9	2.38
11.10	11.10	3.42
11.11	11.11	3.42
11.12	11.12	8.26
11.13	11.13	2.41
	12.1	1.27
12.1	12.3	8.6–8.15, 8.25
12.2	12.4	8.6–8.15
12.3	12.5	6.89, 8.16–8.18
12.4	12.6	8.16–8.18
12.5	12.7	8.20
12.6	12.8	8.22–8.24
12.7	12.9	8.29
12.8		8.29, 8.30, 8.32
	12.10	8.29, 8.34
12.9		8.29–8.33
	12.11	8.29, 8.34
12.10	12.12	8.35
13.1	13.1	9.2
13.2	13.3	9.10
13.3	13.2	9.7
13.4	13.4	9.21
13.5	13.5	9.21

Optional clauses

CBC cl	DBC cl	paragraph
A1	A1	4.8
A1.1	A1.1	4.9
A1.2		4.18
A1.3		4.18
A1.4	A1.2	4.10, 4.19
A1.5	A1.3	4.10
A2	A2	3.12, 3.16, 5.12
A2.1	A2.1	2.26, 5.14

CBC cl	DBC cl	paragraph
A2.2	A2.2	2.27, 5.12
A2.3	A2.3	3.19
A2.4	A2.4	3.12
A2.5	A2.5	3.15
A2.6	A2.6	3.22, 7.12
A2.7	A2.7	3.21, 7.24–7.25
A3	A3	2.17, 4.4, 4.7
A4		6.41
A4.1		6.39
A4.2		6.41–6.47, 6.52
A4.3		6.41–6.47
A4.4		6.41–6.47, 6.66
	A4	6.66
	A4.1	6.66
A5		6.33
A5.1		6.33
A5.2		6.2
A6		6.86
A7	A5	3.26
A7.1	A5.1	3.26–3.29, 5.5
	A6	3.37
	A6.1	3.37
	A7	1.20, 7.3, 2.21
	A7.1	2.21
A8		2.28–2.30, 5.33
A8.1		2.29
A8.2		5.33
A9		1.3
A10	A8	3.44
A10.1	A8.1	3.44
A11	A9	2.31, 4.35, 6.17
A11.1	A9.1	2.31, 4.35, 6.17

Subject Index *by paragraph number*

access, site   4.3, 5.16, 5.38, 8.35
additional payment   6.16, 6.17, 6.30
   rules for valuation   2.31, 4.35, 6.17
adjudication   2.24, 9.10–9.20
advance warning and joint resolution of
      delay   4.22–4.23
advanced payment   6.20, 6.33
Agreement   2.12, 2.37
arbitration   2.24, 9.21–9.25
architect/contract administrator, duties   3.4–3.7,
   5.10–5.11

bills of quantities   2.22
Building Regulations   2.20

calculation of periods   2.41
cancellation by customer   1.27
CDM *see* Construction (Design and
      Management) Regulations (CDM) 2015
certificates
   final payment   6.54, 6.75–6.76
   interim payment   6.21–6.24, 6.44, 6.52,
      6.58–6.59, 6.66
   practical completion   5.36
change to works instructions   2.45, 4.27, 5.31,
      6.13–6.15, 6.16
client   3.36, 3.37
   liability   7.6, 7.8–7.11
   risks   4.26
collateral warranty   2.28–2.30, 5.33
communication and notices   3.42, 8.26
complete works according to contract   3.9,
      4.16
completion in sections   2.17, 2.1, 4.4, 4.7
conclusiveness   6.77
Construction (Design and Management)
      Regulations (CDM) 2015   5.6–5.8, 8.13
consumer   1.23–1.28
Consumer Rights Act 2015   1.28
contract administrator, duties   3.4–3.7, 5.10–5.11
contract documents   2.15–2.16, 2.38
   interpretation   2.39
   priority   2.42
contract execution   2.37–2.38
contract price   2.22–2.23, 6.5–6.6
   adjustments   6.7–6.17
contractor
   application for payment   6.22, 6.65, 6.80, 6.81

indemnity   3.22, 7.12
   liability   7.4–7.5, 7.8–7.11
   obligations   3.9
   right to suspend   6.86–6.88
   risks   4.26
contractor design   2.15, 2.26–2.27, 3.11–3.15,
      3.16, 3.19, 5.12–5.14
contractor's representative   5.4
contracts
   differences   1.12–1.13
   key features   1.3–1.6
cost savings   6.9–6.10
court   9.20
customer acting as contract administrator   3.37

defective work   5.18–5.20, 6.12
defects fixing period   5.46–5.48
deferred access   4.6
definitions   2.40–2.41
delay   4.22–4.23
design   2.26–2.27
design liability, level of   3.16–3.22
dispute resolution   2.24–2.25, 9.1–9.25
documents *see* contract documents
due date for payment   1.18, 6.21, 6.48

early use   5.40–5.44
evidence of ability to pay   1.22, 6.86, 7.23
execution of contract   2.37–2.38
extensions of time *see* revisions of time

facilities   2.13
final contract price   6.53–6.56, 6.74–6.76
final date for payment   6.48, 6.56
final payments
   CBC   6.53–6.56
   DBC   6.74–6.76
fit for purpose liability   2.26, 3.19
fluctuations   1.11, 6.8
force majeure   4.26

health and safety   3.33–3.3, 5.6–5.8
Housing Grants Act 1996   1.14–1.19

improvements   6.9–6.10
inconsistency in the contract documents
      2.43–2.45
inspection   5.16–5.20, 6.11

instructions   2.44–2.4, 4.17, 5.13, 5.19–5.23
   contractor to comply immediately   5.24
   contractor failure to comply   5.30, 6.36
   delivery   5.24–5.28
   response to notice, comply with decision   5.29
insurance   7.13–7.25
   client   2.22–2.23, 7.15–7.16, 7.18, 7.20–7.21
   contractor   7.15, 7.18–21
   evidence of   6.36, 7.23
insurance backed guarantee   2.21
intellectual property   3.15
interest   6.84–6.85
interim payments
   CBC   6.21–6.51
   DBC   6.58–6.72
interpretation of contract documents   2.39–2.41

letters of intent   2.34
liability
   client   7.6, 7.8–7.11
   contractor   7.4–7.5, 7.8–7.11
   design   3.16–3.22
limit of contractor liability, discrepancies   3.11
liquidated damages   2.17–2.19, 5.45, 6.36

mediation   2.24–2.25, 9.7–9.9
milestone payments   6.41–6.47, 6.66

negotiation
   dispute resolution   9.2
   pre-contract   2.5, 2.32–2.36
new building warranty   2.21
non-completion   5.45
notices in writing   3.42, 8.26

occupation before completion   5.40–5.44
oral instructions   5.24–5.28
other appointments   2.14

partial possession   5.41–5.44
party walls   2.20
pay less notice   6.51, 6.70–6.72
payment notice   6.34, 6.79–6.80
payment on completion of the works   6.39
payment on termination   8.29, 8.31–8.34
personal injury   7.5, 7.10–7.11
planning permission   2.20, 3.34
possession, of site   4.2–4.3
practical completion   5.32–5.37
pre-contract negotiations   2.5, 2.32–2.36
pre-start meeting   3.40–3.42
priority of contract documents   2.42–2.43
procurement route   1.7–1.11

products   3.10
professional indemnity insurance   3.21, 7.24–7.25
programme   4.8–4.19
   non-financial consequences for non-
     submission   4.10
   updates   4.18
progress meetings   3.43
public sector clauses   1.3

rate of interest   6.84
regularly and diligently   3.9, 4.4, 4.16, 8.8–8.11
regulations, compliance with   2.20, 3.33–3.35
regulatory consents   2.20, 3.34
repudiation   8.4
required specialists   3.26–3.32, 5.5
retention   5.38–5.39, 5.42, 6.24, 6.32
revisions of time   2.45, 4.24–4.42
   rules for valuation   2.31, 4.35, 6.17
risks register   3.44

schedules of rates or works   2.22–2.23
Scheme for Construction Contracts   9.12–9.17
security   4.2, 4.5
site   2.13, 4.2
specification   2.27
statutory charges   3.34
statutory requirements   3.33
subcontracting   3.23–3.26, 5.5
submission of design details   5.12
suspension   6.86–6.88

tendering   2.3–2.8
termination
   by client   8.6–8.15, 8.25
   by contractor   6.89, 8.16–8.18
   by either party   8.19, 8.29
   due to insolvency   8.2
   payment on termination   8.29–8.34
third party rights   2.30, 5.33

Unfair Terms in Consumer Contracts Regulations
   1999   1.25
updates to contract documents   2.38

VAT invoice   6.49–6.50, 6.67–6.69
valuation   6.25–6.38, 6.61
variation of work see change to works
   instructions

withholding of payments   6.51, 6.70–6.72
working period   2.13
workmanship, contractor's obligations   3.10
written communication   3.42